THE STEPHEN BECHTEL FUND

IMPRINT IN ECOLOGY AND THE ENVIRONMENT

The Stephen Bechtel Fund has

established this imprint to promote

understanding and conservation of

our natural environment.

The publisher gratefully acknowledges the generous contribution to this book provided by the Stephen Bechtel Fund.

North American Amphibians

North American Amphibians

Distribution and Diversity

DAVID M. GREEN, LINDA A. WEIR, GARY S. CASPER, and
MICHAEL J. LANNOO

UNIVERSITY OF CALIFORNIA PRESS

Berkeley   Los Angeles   London

University of California Press, one of the most distinguished university presses in the United States, enriches lives around the world by advancing scholarship in the humanities, social sciences, and natural sciences. Its activities are supported by the UC Press Foundation and by philanthropic contributions from individuals and institutions. For more information, visit www.ucpress.edu.

University of California Press
Berkeley and Los Angeles, California

University of California Press, Ltd.
London, England

Library of Congress Cataloging-in-Publication Data

  North American amphibians : distribution and diversity / David M. Green, Linda A. Weir, Gary S. Casper, and Michael J. Lannoo.
    pages cm
    Includes index.
    ISBN 978-0-520-26672-8 (hardback)
    1. Amphibians—North America.   I. Green, David Martin   II. Weir, Linda A.   III. Casper, Gary S.
  IV. Lannoo, Michael J.
  QL651.N67 2014
  597.8—dc23   2013026611

Manufactured in Hong Kong
22 21 20 19 18 17 16 15 14 13 10 9 8 7 6 5 4 3 2 1

The paper used in this publication meets the minimum requirements of ANSI/NISO Z39.48-1992 (R 2002) (*Permanence of Paper*). ♾

Cover image: *Eurycea spelaea* (Grotto Salamander), upper; *Haideotriton wallacei* (Georgia Blind Salamander), lower. Photos by Dante Fenolio, combined by David M. Green.

The records, notes, observations and field data of hundreds of field biologists constitute the original material that went into the making of this book. To them and their efforts, we dedicate this atlas of North American amphibians.

CONTENTS

PREFACE

This book could only have come to be produced because of the enormous growth in interest about amphibians and the increased intensity of scientific research into their biology and distribution that has occurred over the past two decades. Ironically, this recent surge in the study of amphibians has come about through our realization of a slowly unfolding tragedy. Amphibians are widely recognized now to be in global decline, and a great many species in Canada and the U.S. are endangered or threatened with extinction.

The idea for this book, an atlas of North American amphibians, arose following the publication of two books on the status of amphibians in North America, both of which were written, effectively, as regional reports to the global Declining Amphibian Populations Task Force that have been set up by the International Union for Conservation of Nature (IUCN). The first of these books, *Amphibians in Decline. Canadian Studies of a Global Problem,* was edited by David M. Green and appeared in 1997. Among its contents, this book included a species-by-species account of the known distributions and current status of all 45 frogs, toads and salamanders native to Canada. It was followed by a counterpart for the U.S. in 2005: *Amphibian Declines: The Conservation Status of United States Species,* edited by Michael J. Lannoo. This much larger volume also contained species distribution and status accounts, but in a more comprehensive fashion and with the addition of range maps for each of some 290 species of amphibians native to the U.S. Our initial ambition was to combine the reports and distribution records that went into these two volumes (one Canadian and one American), update the available information, and present a unified accounting of all the amphibians now known in North America north of Mexico.

However, we did not want simply to present range maps such as might be found in a standard field guide. The typical field guide map is intended to provide an impression of where a particular species might be encountered so that it might be more easily identified. A field guide map tends to consist of one or more colored or cross-hatched shapes drawn across a largely featureless outline map. These are sometimes termed "blob" maps because of the form of these shapes, which are intended to encompass all the observational and museum records of a particular species. A good blob map takes into account the landforms and habitats where the species has been found and where it might be encountered and so the map becomes both a description and a prediction of the actual range of a species.

For our atlas, though, we wanted to show more precisely where animals have been found in relation to the topography of the North American continent. Consequently, we set out to plot dot maps against a shaded relief map of North America to demonstrate, in a way that cannot easily be done with a field guide map, how the topography of the continent can shape and direct the actual distributions of species.

The information in the species accounts in this book, with few exceptions, is derived directly from the extensive accounts included by Lannoo in Amphibian Declines. These were comprehensive summations of our knowledge about the species in the U.S. up to 2005, backed by references to the scientific literature and previously unpublished observations. All were written and reviewed by experts on each species and we have retained authors' names with their respective accounts in recognition of their contributions. We have rewritten and updated all of those original accounts and added information pertinent to Canada. We also wrote entirely new accounts for species that had not yet been recognized in 2005. Sources for this new information, as noted in the text, are listed in the Notes section at the back of the book. The species status designations we include are up-to-date according to published and on-line state, provincial, territorial and federal listings in Canada and the U.S. Bear in mind that the scientific names of many species have changed since 2005, although the standard, or common, names have generally remained more stable[1].

The photographs presented in this book are superb images of North American amphibians obtained from a network of our contacts across Canada and the U.S. In most cases, we obtained multiple images of each species and chose the best of them. Of the 297 species of amphibians mapped and discussed in this book, we are able to present photographs of all but the rarest of North American species: the Blanco Blind Salamander.

This book is intended for all people who are aware of and interested in nature, especially those who are interested in what our amphibians look like, where they live, and how they live. Professional biologists interested in other aspects of ecosystems, such as plants or insects, birds, and mammals may also find this book useful. Finally, professional herpetologists will find that this volume serves as a distribution update and photographic supplement to earlier books on North American amphibians.

INTRODUCTION

In profile, North America is an immense plain hemmed by mountains to either side. Along the east are the old, eroded Appalachian Mountains that extend north and east from the southern Coastal Plain to the Gaspé Peninsula, the even more ancient Laurentian Mountains north of the Gulf of St. Lawrence, and the eroded, broken mountains of the Arctic Cordillera that run from northern Labrador and Québec through the Arctic Archipelago to the tip of Ellesmere Island. Along the west, there lie the great spine of the Rocky Mountains and other ranges in parallel rows—a mere section of the vast Cordillera of the Americas that runs from the Beaufort Sea to Tierra del Fuego. To the south, the trough is drained by the Mississippi River into the Gulf of Mexico. In the northeast, the Great Lakes basin drains east into the Atlantic Ocean through the St. Lawrence River, the only major waterway leading directly into the heart of the continent from the east. Further north, rivers from the west drain into the great inland sea of Hudson Bay and, more northerly still, the Mackenzie River leads north to the Arctic Ocean. The greater part of this gigantic land is occupied by just two countries, Canada and the U.S., and in them live nearly 300 species of amphibians.

NORTH AMERICAN AMPHIBIANS

Amphibians are far from being the largest of North America's wildlife. Most are small. Many are tiny. The Little Grass Frog is no more than 1.5 cm end to end and the Two-toed Amphiuma, the longest North American amphibian, may reach only slightly more than a meter in length. Yet their ecologic importance belies their size, for they are, by far, the most numerous terrestrial vertebrates on the continent. Their abundance and diversity also reflect, in a very immediate way, the climate and landscapes of the areas where they are found, and therefore mapping the distribution of amphibians in North America is also an exercise in mapping the climate and topography of the continent itself.

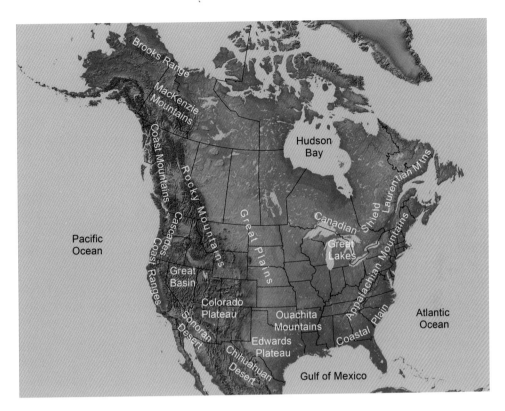

North America, physical.

Two of the three living orders of amphibians are to be found in Canada and the U.S. Frogs, toads and treefrogs, the Order Anura, are distributed throughout the world except for the polar regions and remote oceanic islands. Salamanders and newts, the Order Caudata, are distributed mainly in the Northern Hemisphere and the New World Tropics. Missing from Canada and the U.S. are any caecilians, Order Gymnophiona, which are strictly tropical.

More than anything else, amphibians' thin, moist, highly vascularized and water-permeable skin sets them apart from other terrestrial vertebrate animals. The skin of amphibians has no epidermal coverings such as fur, scales, or feathers and, with few exceptions, is not waterproof. Amphibians are rarely found far from moisture, as their permeable skins do not well protect them from desiccation in dry conditions. Amphibians also cannot withstand the salinity of seawater, which draws away their body water by osmosis. However, amphibians do not necessarily require standing water because they can absorb the water they need through their skin, even from the ground or from moist soil. This is the ability that allows desert amphibians to survive underground, where surface environments are intensely dry, and enables woodland salamanders to thrive far from standing water.

Amphibians also exchange respiratory gasses through their skin, in addition to their lungs. Cutaneous gas exchange enables many northern amphibians to hibernate under water in winter, as they can receive enough oxygen for respiration through their skins without having to surface for air. Plethodontid salamanders have dispensed with lungs entirely and rely only upon respiration through the skin and through the lining of the mouth. Being able to respire and draw water through the skin enables amphibians to prosper in both aquatic and terrestrial environments and to flourish in dampness.

Amphibians are notable for their skin glands. Small mucous glands distributed all over the body secrete fluid mucus that helps lubricate the skin and keep it from drying. Mucus is also secreted when the animal is stressed, helping to make it slippery and difficult for predators to capture or hold. The granular glands of amphibian skin are larger than the mucous glands and are located in discrete places on the upper surface of the body, tail, and head. They secrete toxic and noxious substances, primarily for defense. No amphibian has a poisonous bite, nor, with rare exceptions, can an amphibian inflict a poisonous wound. The granular gland secretions are released when the animal is in duress, mainly in order to render itself a less than tasty meal. Nonpoisonous species have small granular glands that release few toxic compounds, but poisonous species may secrete substantial amounts of bitter-tasting, irritating, noxious, and even highly toxic nitrogen-containing chemical compounds. The warts and parotid glands of toads and the granular bumps on the skin of newts are visible concentrations of granular skin glands.

Amphibians are ectotherms, whose body temperatures are in near-equilibrium with their surroundings. This means they are able to be active only at moderate temperatures, above freezing. By virtue of being ectotherms, though, amphibians do not metabolize their food to produce bodily heat. Consequently, they eat much less and use much more of their intake for growth and reproduction, by roughly an order of magnitude, compared to similarly sized mammals. Amphibians are environmental opportunists capable of rapid growth and considerable investment in reproduction under the right thermal conditions. Their metabolic inability to maintain internal body temperatures has not, however, prevented certain amphibians from surviving hot desert summers and or cold northern winters. The species that live in these habitats endure by escaping underground and by having adaptations that allow them to remain dormant until surface conditions are once more to their advantage.

When an amphibian is very cold, with its body temperature hovering just above freezing, its metabolic activities slow to a minimum. Oxygen demand becomes so low that enough can be obtained entirely through the skin, and though the animal is too cold and slow to feed, its nutritional requirements are so minute that it does not need to; it relies instead on stored energy reserves in the form of fat and glycogen. In this condition, amphibians can remain dormant throughout the winter, so long as they can remain in a place where they are protected. American Bullfrogs, Green Frogs, Northern Leopard Frogs and Mink Frogs hibernate underwater well below the surface ice. These frogs, when cold, can gain enough oxygen through their skins without needing to breathe air. Western Toads, American Toads and Eastern Red-backed Salamanders, among other northern species, seek refuges or bury themselves below ground to avoid frost. For them, as for most animals, freezing is deadly. But a few species, such as Wood Frogs, Spring Peepers, Boreal Chorus Frogs, and Gray Treefrogs, do not need to avoid freezing; they have antifreeze. As they cool, they flood their blood systems with glucose or glycerol; these substances enable them to tolerate substantial periodic freezing of their body water and still survive. When temperatures rise in the spring, the metabolic activity of the hibernating amphibian will also rise, and it will again become active. During mild winters on the Pacific Coast or southeast of the continent, the temperature may never get very low for very long and some species of amphibians may remain active all winter.

Amphibians may also enter into extended periods of dormancy during times of severe heat and dryness. This is called "aestivation," during which an amphibian will greatly reduce its activity and metabolic rate and remain burrowed underground, sometimes for years. Spadefoots survive in deserts owing to their ability to aestivate and dig down in sandy soils to where they can absorb soil moisture and remain cool. Sirens can survive periods of drought by aestivating in dried mud and forming a cocoon of dried skin around themselves.

In the spring, the voices of frogs fill the night air over most of North America. Vocalization is the primary means of communication for almost all frogs. Each species has its own distinct repertoire of calls, generally given only by males. Male Gray Treefrogs and Cope's Gray Treefrogs, which are otherwise virtually identical, can be readily distinguished in the wild by their calls. Among North American species, only the two species of tailed frogs make no vocalizations. The songs of male frogs are produced by the passage of air rapidly back and forth over the vocal chords between the lungs and the buccal cavity. The mouth and the nostrils are kept closed while the frog is calling, enabling some species, such as Columbia Spotted Frogs, to call while under water. Most species of frogs have expandable vocal sacs that serve to increase the volume of sound and radiate it outward. With few exceptions, salamanders are silent and identify each other by scent rather than sound.

All species of North American amphibians lay eggs. Fertilization of the eggs may be either external, as in Hellbenders and most frogs, or internal, as in the tailed frogs and most salamanders. With external fertilization, the male applies sperm to the eggs as they are being laid. When frogs mate, the male clasps the female with his forearms in an embrace called amplexus. Tailed frog and spadefoot males grasp their partners around the waist. Males of other species clasp the female behind her forearms. Among frogs, only tailed frogs employ internal fertilization, and among amphibians, only tailed frog males possess a copulatory organ. Most salamanders practice internal fertilization by means of a sperm capsule, called a "spermatophore," which the male lays and the female picks up with her cloaca. This procedure often involves an elaborate mating ritual that enables salamanders of the same species to recognize each other, and position themselves so that successful sperm transfer can occur.

Amphibian eggs do not have hard protective shells, and the embryos do not grow fluid-holding external membranes. They are surrounded only by a transparent jelly coat and must either be laid in the water or be hidden in moist places on land to protect them from drying. Most North American amphibians lay aquatic eggs and hatch as tiny, gilled, tadpoles or larvae. Many species of frogs will lay thousands of eggs at a time. The larval stage may last from several weeks to several years, depending on the species, before the animal metamorphoses

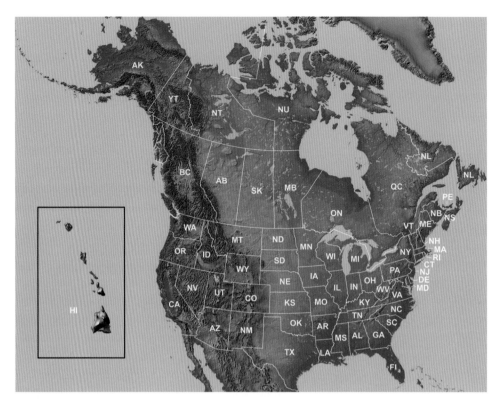

North America and Hawaii, political, showing U.S. states and Canadian provinces.

into its adult form. Some species of salamanders, though, retain their larval appearance their entire life and will reach sexual maturity while still in larval form. This condition is called "neoteny" and is characteristic of sirens, Mudpuppies and many cave-dwelling species. In some populations of Northwestern Salamanders, tiger salamanders, and giant salamanders, individuals are also frequently neotenic. On the other hand, many terrestrial plethodontid salamanders, such as Ensatinas slender salamanders, and slimy salamanders, and many tropical and subtropical frogs, including the chirping frogs and robber frogs, bypass the free-living larval stage by going through direct development to the adult form while within the egg. The young of these species hatch as tiny replicas of their parents. Parental care is commonplace among these direct-developing species, which all lay their eggs in moist places on land. Either the female or the male parent tends to the eggs, keeps them moist with its own body, and eliminates dead or diseased eggs by eating them. If a tending adult is removed from a clutch of terrestrial amphibian eggs, the eggs have a much-reduced chance of survival.

Amphibians play an important role in the energy flow of ecosystems as small, abundant predators and aquatic herbivores and detritivores. The tadpoles of frogs consume aquatic plants, detritus, and carrion, but when they grow and transform into adults they become carnivores. Salamanders and their larvae are strictly carnivorous throughout life. Amphibians consume large numbers of insects, worms, or other invertebrates and, in turn, serve as a food source for snakes, raccoons, birds, or other larger predators. The great abundance of amphibians thus forms an important link in the food chain.

No amphibian is directly harmful or dangerous to humans, but humans, directly and indirectly, are having an enormous impact on amphibians. Recent indications of widespread amphibian population declines have raised great concern about North American amphibians and greater awareness of the difficulties they may face in a world where the human footprint is increasingly heavy. Habitat loss due to land clearing for urban development, agriculture, logging, and other reasons, is a major threat to amphibians. Water development and drainage projects such as dams or canals can destroy habitat by scouring stream banks, inundating habitats, eliminating ponds and marshes, and causing drought during critical periods of amphibian development. Other problems for amphibians result from introduced animals, including stocked fish, which may be brought in for commercial or recreational purposes without regard to their ability to outcompete native animals, feed on them, or transmit diseases. Human disturbance of streams and ponds can lead to increased siltation that can smother eggs and larvae. Increasing automobile traffic and roads result in an increased incidence of road kills and fragmentation of habitats. Amphibians are also disappearing from near-pristine areas where there is no visible disturbance. Chemical contamination from acid rain and contaminated snowmelt runoff, heavy metals leaching from mining operations, fertilizers and pesticides, and chemicals used in road maintenance and construction may harm amphibians directly or cause a host of sublethal effects on the animals' health. No cause is more insidious in the demise of populations of some North American frogs than an emerging disease caused by a recently recognized chytrid fungus. Though declines and extinctions occur naturally in all animal populations, in recent years there have been both declines in amphibian abundance and losses of amphibian populations altogether.

THE NORTH AMERICAN CONTINENT

The geologic formation of North America dates back nearly 2 billion years, when the ancient, stable crust of the Canadian Shield coalesced from four disparate continental fragments, or cratons, and fused with the equally ancient Precambrian crust that underlies the present-day northern Great Plains. These vast, flat expanses of ancient rock are exposed in the northeast, where they were scraped clean by the comparatively recent continental ice sheets of the Pleistocene, beginning a mere 2.5 million years ago, but toward the south they are overlaid with layer upon layer of sedimentary rocks dating back over 600 million years.

Between 600 million and 250 million years ago, this early North America was located close to the equator, and much of it was submerged. In these shallow seas, ancient corals grew and their remains resulted in thick accumulations of limestone. Because limestone will slowly dissolve in water, over time such limestone, or karst, regions become riddled with subterranean passages, caves and water-filled aquifers. This is the origin of the underground waters and springs of the southeastern U.S., in which many species of cave-dwelling and wholly subterranean salamanders now live.

The mountains that surround the central plain of the continent first began to rise 400 million to 300 million years ago, when North America collided with other continents to form the ancient supercontinent of Pangaea. Where the northeast-facing margin collided with northwestern Europe, the Arctic Cordillera took shape. Where the south-facing margin collided with South America, there formed the Ouachita and Ozark Mountains. And where the southeast-facing margin collided with northwestern Africa, ridge after ridge of the Appalachians rose, including the Blue Ridge Mountains to the southeast, the Cumberland, Allegheny, and Adirondack Mountains to the northwest, and the Appalachian Valley and Ridge region in between. All of these ancient mountain ranges have been well worn down by now and the various ridgetops of the Ozarks, Ouachitas and Appalachians today are rich in species of amphibians, particularly salamanders.

The mountain ranges that formed along the western margin of the continent are much younger, and higher, than the Ozarks, Ouachitas, and Appalachians. About 180 million years ago, North America began to drift westward over the floor of the Pacific Ocean. As it went, it sheared rocks off the Pacific Plate, which accreted in successive ridges onto its western edge. The process also resulted in chains of volcanoes just inland of the western margin of the continent. This process continues today, resulting in the great chain of upthrust sedimentary rocks known as the Rocky Mountains and the arc of coastal volcanoes stretching from the Aleutian Islands through the British Columbia Coast Mountains and the Cascades Range to the Sierra Nevada of California. In between these two long mountain chains lies a region of stretched and crumpled continental crust know as the Basin and Range Province. These comparatively new, dry, and rugged regions are home to only a small number of species of amphibians.

The westward drift of North America that created the Rocky Mountains and coastal ranges also created the Atlantic Ocean and the Gulf of Mexico as Pangaea split apart. This has given the continent its low-lying Atlantic and Gulf Coastal Plains, interrupted by the Mississippi River. Many species of amphibians, especially frogs, have overlapping ranges along these coastal plains such that more species can be found living in the same place than in any other region of the continent.

GLACIATION

The primeval origins of North America's plains, mountains, waterways, and valleys set the stage for the animals and plants living here today. But the current distributions of amphibians in North America are profoundly related to geologic events much more recent than the accretion of cratons, the drift of the continent, and the rise of its mountains. Beginning about 2.5 million years ago, northern North America entered a cycle of periodic episodes of glaciations, when colossal ice sheets covered half the continent. The most recent glacial maximum was about 18,000 years ago. At that time, ice sheets as much as 4 kilometers thick spread from the Laurentian Mountains and the Canadian Cordillera to cover virtually all of present-day Canada and the Great Lakes basin. So much of the earth's water was frozen into continental ice sheets that the global mean sea level was 120 meters lower than

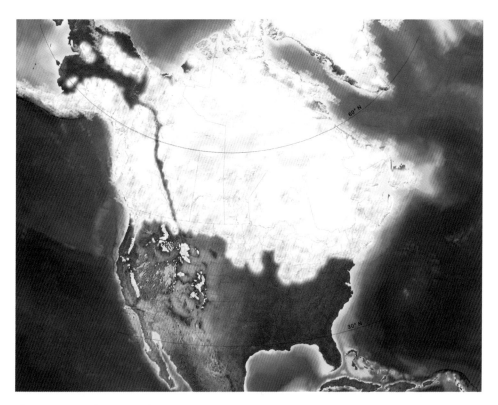

Pleistocene North America during the most recent Wisconsinan glacial maximum, 12,000 to 15,000 yrs ago. Image copyright Ron Blakey, Colorado Plateau Geosystems, used with permission.

it is today. The recession of this blanket of ice took place from about 13,000 to 6,000 years ago, leaving behind easily visible signs of the immense power of the ice sheets. The movement of the ice scoured the Canadian Shield to leave smooth, bare bedrock surfaces and ice-cut rock basins that are now filled by innumerable lakes. The debris carried off by the glaciers was deposited to the south, most visibly as a great terminal moraine, a line of hills marking the furthest extent of the ice. Once the tremendous weight of the ice had been lifted from the crust, the land began to rise, and it is still rising to this day. Sand hills on the coastal plain were once beaches by the sea; they are now stranded inland as the plain slowly rose over thousands of years. Meltwaters of the receding ice formed immense, but ephemeral, proglacial lakes. Surrounding each of the modern remnants of these lakes— Lake Winnipeg, Lake Athabasca, Lake Champlain, and many others—are similar lines of sand hills that mark their ancient beaches. Salt flats throughout today's western deserts are the dried remains of Pleistocene lakes.

No amphibian could possibly survive where there was such ice. Consequently, all amphibian populations currently living in previously glaciated North America arrived there within the past 6,000 years or so and all are descended from populations that had survived south of the ice. However, they did not necessarily need to find refuge very far south of the ice.

Though the climate was undeniably cooler during a glacial episode than it is during today's interglacial period, it need not have been profoundly cold. All that was necessary for the ice sheets to form was for the climate to be cold enough at their centers of origin for winter snows to remain unmelted during summer. Each succeeding winter, more and more snow could then accumulate. The compressed snow became ice, and the thick ice began to flow. The kilometers-thick ice sheets of the Pleistocene flowed far to the south from their centers of origin, but the cool climate that made them remained in the north. During the time of the glacial maximum, it is probable that many northerly distributed species of amphibian—such as Wood Frogs, Spotted Salamanders, and Western Toads—survived just kilometers away from the ice.

To the south of these ice sheets, the ranges of widespread amphibians were more restricted than they are today, and many glacial-era refuges were located on either side of mountain ranges, particularly the Appalachians and Pacific Coast Ranges. The repeated range restrictions caused by the advance and retreat of the ice sheets are the origins of different genetic lineages observed within many species today, including divisions between closely related species. This has resulted, for instance, in the differentiation of Coastal Tailed Frogs, Coastal Giant Salamanders, and Van Dyke's Salamanders, located in the Cascades Range and Coast Mountains, from their close relatives, Rocky Mountain Tailed Frogs, Idaho Giant Salamanders, and Coeur d'Alene Salamanders, found in the Rocky Mountains. The northern shift of ranges following the retreat of the ice also left behind isolated populations of amphibians in shrinking patches of suitable habitat. These glacial relicts include Wyoming Toads, Houston Toads, Vegas Valley Leopard Frogs, various species of slender salamanders, and others.

Despite their intolerance of seawater, amphibians today inhabit many offshore islands. The presence of toads, frogs, and salamanders on Vancouver Island, the California Channel Islands, the Barrier Islands of North Carolina, Long Island, and Martha's Vineyard indicates that overland connections to the mainland existed during the time of Pleistocene glaciation. No amphibians are native to the Island of Newfoundland or Anticosti Island, though, as these islands became separated from the mainland before amphibians could reach them following the retreat of the ice.

CLIMATE AND AMPHIBIAN DISTRIBUTION

Amphibians, attuned as they are to temperature and moisture, are profoundly influenced by climate. North America is enormously wide and is located at latitudes where most weather systems move from west to east, propelled by the upper jet stream that snakes across the continent. In general, North American weather systems reach the West Coast laden with moisture picked up from the Pacific Ocean. As these wet air masses rise over the Coast Ranges along the Pacific Coast, they cool and lose their moisture as precipitation. This results in the warm, wet weather and moderate climate of the maritime west coast forests, which are dominated by the largest trees in North America. The gigantic redwoods and western red cedars that grow in these forests along the north coast of California are replaced by Douglas firs and western hemlocks in Oregon, Washington, and British Columbia that are almost as large. Rough-skinned Newts, the torrent salamanders, Northern Red-legged Frogs, and Coastal Tailed Frogs flourish in this environment. To the south, in coastal southern California, the ameliorating effects of the wet westerly winds are confined largely to wintertime. The summers are hot and arid, under the influence of dry air masses that expand out from the desert inland. This is a Mediterranean climate that fosters an open scrub forest of chaparral, evergreen oaks, yellow pine, and sagebrush. In this region live the slender salamanders, Arboreal Salamanders, Western Spadefoots, and Arroyo Toads.

As the westerly winds crest the mountains and descend into the western intermontane basins and valleys, they have been wrung free of most of their moisture. In this shadow of the mountains, from the Okanagan Valley of south-central British Columbia through the Great Basin to the arid lands of Arizona and New Mexico, there is desert. Drought-tolerant plants as sagebrush and bunchgrasses cover the high deserts toward the north, whereas the low-lying southerly deserts feature heat-tolerant yucca, saltbush, mesquite, and cacti of many varieties. This is a harsh environment for amphibians, but Great Basin Spadefoots, Arizona Toads, and Canyon Treefrogs make their homes here.

The high inland mountains in the northwest also force the weather systems further upward, resulting in an interior wet belt in more northerly latitudes. These northwestern mountain forests are made up of cedar and Douglas fir on the wet western slopes, lodgepole pine and white spruce on the dry eastern slopes, and Sitka and Engelmann spruce at higher elevations up to the tree line. In these forests, there are Western Toads, Columbia Spotted Frogs, Coeur d'Alene Salamanders, and Long-toed Salamanders.

Once across the Rockies, the westerly air masses descend again over the Great Plains and prairies of the continent's central and southern interior. These vast grasslands fill the midsection of the continent from the Canadian prairie provinces of Alberta, Saskatchewan, and Manitoba south to Texas, separating North America's western mountains from its eastern forests. The dry, western Great Plains are covered by short grass prairie vegetation consisting mainly of bunchgrasses and sagebrush, but in the river valleys and coulees cut by smaller streams, more luxurious vegetation and trees can grow. The eastern plains are more humid, and its rich soil supports a dense tall grass prairie of bluestem grasses, indiangrass, and switchgrass, along with a profusion of wildflowers and berry bushes. The Great Plains are inhospitable for most amphibians. Plains Spadefoots, Great Plains Toads, Canadian Toads, and Plains Leopard Frogs are found associated with coulees and prairie pothole ponds but, except for the blind, cavern-dwelling salamanders of the Edwards Aquifer in west-central Texas, the only salamander of the plains is the Western Tiger Salamander.

In the southeast of the continent, to the east of the Great Plains, the westerly winds once again bring moisture as warm, humid air moves up from the south in spring and summer from off the Gulf of Mexico. This combination brings rains and a moderated climate to the Appalachian Mountains and their surrounding interior and coastal plains, and fosters the growth of the great eastern temperate forests. These mixed forests contain both

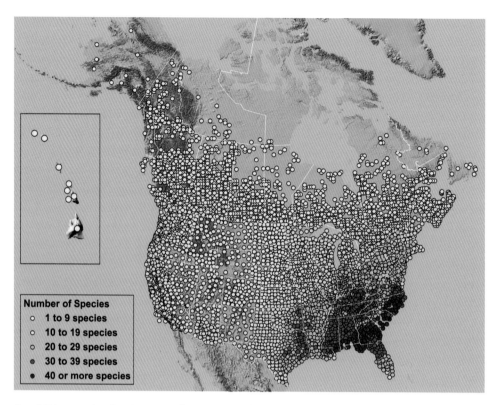

Amphibian species density in North America and Hawaii.

hardwood deciduous trees, such as maples, beeches, oaks, hickories, poplars and birches, and softwood coniferous trees, such as pines and hemlocks. Toward the south, the forests increasingly contain more subtropical trees, such as pawpaws, myrtles, magnolias, laburnums, mimosas, live oaks, and gum trees. Regions of dry, sandy soils that are a legacy of glacial-era drainage patterns are covered by pines. The rich, humid conditions of the eastern forests are ideal for amphibians, and it is in these regions where most of North America's amphibian species are found. The lowlands support a profusion of mole salamanders, amphiumas, waterdogs, treefrogs, gopher frogs, chorus frogs, and cricket frogs. On the Atlantic and Gulf Coastal Plains, more amphibian species can be found in any one place than anywhere else on the continent. The eastern, forested mountains, though, are the empire of salamanders. On virtually every ridge and mountain crest there will be at least one species of small woodland salamander such as a Northern Zigzag Salamander, a large woodland salamander like Jordan's Salamander, a brook salamander such as the Southern Two-lined Salamander, a species of slimy salamander, and several species of dusky salamanders of varying sizes, along with an assortment of frogs, toads, treefrogs, mole salamanders, and newts.

The only truly tropical wet forests in North America are the hardwood hammocks of southern Florida. These small patches of tropical vegetation, featuring strangler figs, mahogany trees, palms, gumbo limbo trees, bromeliads, and orchids, are scattered throughout the marshland of the Everglades like islands of rainforest in a sea of sawgrass. Despite their lushness, these hammocks have few amphibians, mainly treefrogs. Various aquatic amphibians, such as dwarf sirens, Eastern Newts, and Pig Frogs live in the surrounding marshes.

In the north of the continent, the influence of the westerly winds is weak. Instead, the major climatic force is the cold, dry air of the Arctic, which descends south in winter, particularly through the flat, central interior of the continent. In the plains of the far north and northeast, where the arctic air supplies only a light coat of winter snow and a meager fall of summer rain, there is tundra. The intense cold of the winter and shortness of the summer in these regions keeps surface water from easily draining over the flat land. The result is a marshy plain, underlaid with permafrost, where trees cannot grow. Amphibians, too, are stymied by the tundra, except for Wood Frogs, which are able to venture out onto its southern rim. Just south of the tundra, though, and crossing the continent in a great, unbroken swath from the Bering Sea to the Atlantic Coast of Canada, is the boreal forest of North America. This is a forest of conifers, including spruce and balsam fir, jack pine, and tamarack, and all of it is on previously glaciated terrain. Like the forest, amphibians such as Spring Peepers, American Toads, Mink Frogs, Wood Frogs, and Blue-spotted Salamanders that inhabit North America's boreal region are new arrivals following the retreat of the Pleistocene ice sheets.

Humans are the most recent influence on the distribution of North American amphibians. Many species have been introduced to localities outside their native range, sometimes deliberately. American Toads and Northern Leopard Frogs were introduced into Newfoundland. Pacific Treefrogs were brought to the Queen Charlotte Islands from the adjacent mainland by someone who liked to hear them sing. American Bullfrogs and Green Frogs were introduced in many parts of western North America in the early 20th century as stock for frog farms. The frog farms failed, but the frogs remain and have since become well-established invasive species. Wandering Salamanders may have come to Vancouver Island in the late 19th century as stowaways on oak-bark shipments from northern California. Seal Salamanders, Shovel-nosed Salamanders, and Black-bellied Salamanders, marketed as "spring lizards" and sold as fish bait, now occur in many localities outside their natural ranges. Cane Toads, Coquís, Green and Black Dart-Poison Frogs, and Greenhouse Frogs were introduced on many of the Hawaiian Islands and in southern Florida, where they are well established and even considered to be pests.

FROGS OF NORTH AMERICA

Readers seeking supporting citations for the frog species accounts in this section may turn to the Notes section at the back of this book, as well as the original, unabridged species accounts with extensive citations on pages 382–600 in *Amphibian Declines,* from which most of the information in these accounts is derived.

Ascaphus montanus Nielson, Lohman, and Sullivan, 2001

Rocky Mountain Tailed Frog

Rocky Mountain Tailed Frogs occur in southeast Washington and northeastern Oregon, in west central and northern Idaho, in southeastern British Columbia and in western Montana. The population in the Columbia Mountains of British Columbia appears to be isolated from the rest of the range. Rocky Mountain Tailed Frogs are found up near to timberline, which is at an elevation of 2100 m in the Wallowa Mountains.

Both adult Rocky Mountain Tailed Frogs and their tadpoles occupy cold, swift mountain streams with cobble substrates. Tadpoles typically require permanent water. Adult Rocky Mountain Tailed Frogs seldom move more than 10 m upstream or downstream, although they may move preferentially into smaller, more shaded streams during the summer. After heavy rains or dews, especially during the spring and autumn, adults and juveniles may be found on land in moist woods.

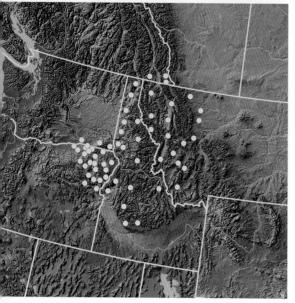

Rocky Mountain Tailed Frog

The specialized habitat of Rocky Mountain Tailed Frogs makes them vulnerable to habitat degradation and destruction following timber harvest or road construction near their habitats. Sedimentation and warmer water temperatures in streams tend to be associated with lower abundances of Rocky Mountain Tailed Frogs.

In Canada, Rocky Mountain Tailed Frogs are considered to be endangered, and they are on the British Columbia Red List of Threatened and Endangered Species.

ORIGINAL ACCOUNT by Michael J. Adams; photograph by Gary Nafis.

Ascaphus truei Stejneger, 1899

Coastal Tailed Frog

Coastal Tailed Frogs occur from sea level to near the timberline in the Coast Mountains of British Columbia north to the Nass River, in the Cascades Range, in the Olympic Mountains, and in the Coast Range in the U.S. south to the Mendocino Range in California. This range is divided by fjords, deep valleys, and major rivers such as the Columbia and Fraser, which limit the dispersal of Coastal Tailed Frogs.

Coastal Tailed Frogs occupy cold, swift mountain streams with cobble substrates. Their tadpoles typically require permanently running streams. Severe floods that scour stream bottoms can remove entire tadpole populations. Adults may move into smaller, more shaded streams during the summer, as they avoid warm water temperatures. After heavy rains or dews, adult Coastal Tailed Frogs may be found on land in moist woods, especially during the spring and autumn.

Lower abundances of Coastal Tailed Frogs have been documented following timber harvest and local road construction. They have been characterized as both environmentally sensitive and resilient to large-scale disturbance. They were one of the first vertebrates to recover following the 1980 eruption of Mt. St. Helens.

Coastal Tailed Frogs are listed a species of Special Concern under the Canadian Species at Risk Act and are on British Columbia's Blue List of species of Special Concern.

ORIGINAL ACCOUNT by Michael J. Adams and Christopher A. Pearl; photograph by Gary Nafis.

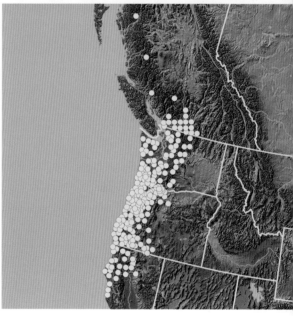

Coastal Tailed Frog

Anaxyrus americanus Holbrook, 1836

American Toad

American Toads are found throughout most of North America east of the central U.S. plains and Canadian prairies. Their northern limit goes from north of Lake Winnipeg to Hudson Bay at the Winisk River in Ontario and the Great Whale River in Québec and east to Lake Melville on the Labrador coast. To the south, American toads are absent from much of the Gulf and Atlantic Coastal Plains, including Florida, except for an extension south through eastern Mississippi and eastern Lousiana to about Lake Pontchartrain. They are absent from eastern Long Island, and are introduced on Anticosti Island and the island of Newfoundland. American Toads are found at elevations over 1524 m in the Great Smoky Mountains.

American Toads are to be found in a great range of habitats, including gardens, fields, lawns, and barnyards. Adults congregate to breed in shallow, often grassy, areas within lakes, ponds, streams, and assorted types of ephemeral wetlands, preferentially not overly shaded by trees. American Toads are mainly nocturnal and seek cover during the day under stones, boards, woodpiles, walkways, porches, or other cover. These toads establish small home ranges of 6 m² or more in size and will repeatedly use particular hiding places. During wet periods, American Toads may travel great distances, easily exceeding 1 km. They are tolerant of brackish estuarine waters.

American Toad

American Toads are considered the common and abundant "hoptoad" of the East. They tolerate humans well and are frequently abundant throughout their range. They do not appear to be as sensitive to habitat fragmentation as are many other species of co-occurring amphibians, and they are often the first amphibian to reinvade clear-cut or burned areas of forest. Tadpoles are, however, susceptible to the low pH caused by acid precipitation and chemical contaminants.

Nowhere is the American Toad considered to be under any threat of disappearance, although population sizes can fluctuate dramatically.

ORIGINAL ACCOUNT by David M. Green; photograph by Brad Moon.

Anaxyrus baxteri Porter, 1968

Wyoming Toad

Wyoming Toads are native only to the flood plains of the Big and Little Laramie rivers at an elevation of 2164 m in the Laramie Basin of Wyoming.

Wyoming Toads frequent a variety of wetland habitats in shortgrass prairie, including lakes, ponds, streams, marshes, and roadside ditches. Adults live close to water and are almost always restricted to the shoreline. Wyoming Toads deposit their eggs in shallow areas of ponds and small lakes, and metamorphosis usually takes place in early July, which corresponds to the annual bloom of small black flies, possibly a food source for young toads. Wyoming Toads have been found near pocket gopher or ground squirrel burrows in the spring and fall, and it is assumed that these are used for hibernation.

The abundance of Wyoming Toads has greatly declined, and they have disappeared from most of their former ranges. The only known current localities in nature for Wyoming Toads are in and around Mortenson Lake. Considering their limited numbers, low genetic diversity, reliance on captive breeding, and continuing problems with disease, Wyoming Toads are likely the most endangered amphibian species in North America.

Wyoming Toads are listed as Endangered under the U.S. Endangered Species Act.

ORIGINAL ACCOUNT by R. Andrew Odum and Paul Stephen Corn; photograph by Sarah Armstrong.

Wyoming Toad

Anaxyrus boreas Baird and Girard, 1852

Western Toad

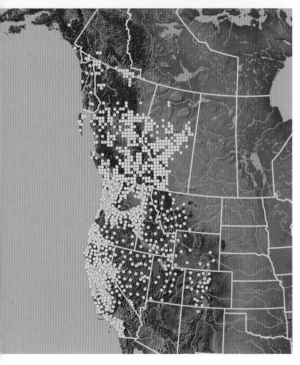

Western Toad

Western Toads occur throughout most of the mountainous regions of western North America. Their range encompasses the entire Pacific Coast from Prince William Sound in southern Alaska to extreme northern Baja California, including Haida Gwaii and Vancouver Island. To the east, they can be found to the Rocky Mountain foothills, including north central New Mexico and western Colorado, Wyoming, and Montana, and western and central Alberta. Western toads are also found in far northern British Columbia, southern Yukon, and extreme southwestern Northwest Territories, generally associated with warm springs in that region. They can occur at elevations to nearly 3400 m in western Wyoming.

Adult Western Toads breed in wetlands in still or barely moving water, typically ponds and small lakes, streams, rain pools, and ditches. Juveniles move from their natal wetlands to nearby terrestrial sites or to other nearby wetlands. Adult Western Toads are often found at the water's edge or basking on partially submerged logs in the spring and early summer. Later in the year, they use a variety of habitat types from upland aspen/conifer stands to rocky areas. During cold weather, Western Toads use gopher and ground squirrel holes as retreats. At higher elevations, Western Toads hibernate in rock-lined chambers near creeks, in ground squirrel burrows, under the root systems of evergreen trees, under large boulders, and in beaver and muskrat lodges.

Severe declines and extirpations of many populations of Western Toads have occurred in areas where they once were abundant, especially in the Rocky Mountains of Colorado and Wyoming. In New Mexico, a number of Western Toad populations have declined rapidly and are now thought to be extirpated. Western Toads are currently abundant within the blast zone of the 1980 Mount St. Helens eruption; however, they are rare in the lowlands of Puget Sound in Washington and the Fraser Valley of British Columbia. Habitat degradation and destruction, including fish introductions and disturbances due to livestock, fungal infections and other pathogens, acid and mineral pollution from mine water drainage, and increased ultraviolet radiation, is contributing most to the decline of Western Toads.

The U.S. Fish and Wildlife Service lists Western Toads in Colorado, New Mexico, and Wyoming as a candidate species for listing. The Colorado Division of Wildlife and the New Mexico Department of Game and Fish list Western Toads as Endangered, and Western Toads are a Protected Species in Wyoming. The Utah Division of Wildlife Resources, as well as the U.S. Bureau of Land Management, has listed them as Sensitive in Utah. Western Toads are listed as a species of Special Concern under the Canadian Species at Risk Act, though British Columbia considers them not to be at risk.

ORIGINAL ACCOUNT by Erin Muths and Priya Nanjappa; photograph by Darlene Hecnar.

Anaxyrus californicus Camp, 1915

Arroyo Toad

Arroyo Toads occur in southwestern California and adjacent northern Baja California in Mexico, (not shown) mostly on the coastal slopes. They also occupy a few drainages on the desert slopes of the San Gabriel and San Bernardino ranges in California.

Arroyo Toads are closely associated with low-gradient drainages that have extensive terrace systems, braided channels, and large areas of fine sediment deposits that are episodically reworked by flooding. They construct shallow burrows within the riparian zone, where they shelter by day during their active season.

Arroyo Toads have declined in abundance, often to extirpation, at most historical sites. Surviving populations have suffered because of drought, development, and adverse land-use and water-management practices, which include the widespread alteration of the middle reaches of larger drainages by dams and flood control projects. These practices have highly fragmented the current distribution of Arroyo Toads through the loss of coastal lowland habitats. Many isolated populations are either restricted to small headwater drainages above impoundments, where conditions are marginal, or confined to narrow riparian corridors along larger drainages that are subject to extensive disturbance from water-management practices, gravel mining, urbanization, and military training.

Arroyo Toads are listed in the U.S. as Endangered.

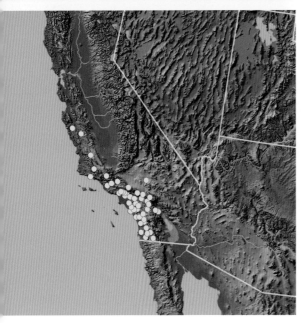

Arroyo Toad

ORIGINAL ACCOUNT by Samuel S. Sweet and Brian K. Sullivan; photograph by Gary Nafis.

Anaxyrus canorus Camp, 1916

Yosemite Toad

Yosemite Toads are restricted to high elevations, from about 2000 to 3500 m, in the central Sierra Nevada Mountains of California from Ebbetts Pass to the Spanish Mountain area.

Yosemite Toads inhabit high-elevation, open, montane meadows; willow thickets; and adjoining forests. Although adult Yosemite Toads spend little time in water, they are seldom found more than 100 m away from permanent lakes, ponds, or streams. During the day, adults take cover in rodent burrows, under surface objects, and in willow thickets.

Populations of Yosemite Toads have disappeared from about half of their former range, largely at lower elevations on the western edge of their distribution. Yosemite Toads may have declined in abundance because of disease, drought, or airborne contaminants; livestock grazing may also contribute to their decline because of trampling, alteration of meadow habitat, and possible lowered water quality. Yosemite Toads may also be subject to increased predation by introduced stocked fish and by increasing numbers of Common Ravens.

The State of California lists Yosemite Toads as a Species of Special Concern. The U.S. Fish and Wildlife Service considers Yosemite Toads a candidate species under the Endangered Species Act.

ORIGINAL ACCOUNT by Carlos Davidson and Gary M. Fellers; photograph by Joyce Gross.

Yosemite Toad

Anaxyrus cognatus Say, 1823

Great Plains Toad

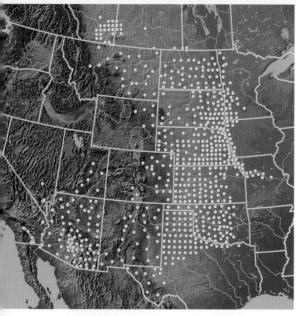

Great Plains Toad

Great Plains Toads occur across the southern Canadian Prairies and the Great Plains of the U.S. from southeast Alberta, southern Saskatchewan, and eastern Montana to the southwestern corner of Manitoba and western Minnesota; south to western Oklahoma and northern Texas; west to the Imperial Valley of California and up the Colorado River though eastern Nevada and southern Utah. Their range extends south to central Mexico (not shown). Great Plains Toads generally are found at elevations less than 1900 m but go up to near 2500 m in the San Luis Valley of Colorado.

Great Plains Toads are found in shortgrass and tallgrass prairies, sandhills, desert mesquite, or desert scrub, but rarely in upland woodlands. They are associated with temporary ponds, irrigation ditches, and bottom lands. Great Plains Toads are tolerant of agriculture and urban conditions. They are proficient burrowers and will form deep burrows in the shape of an inverted question mark, with the toad positioned at the upper, terminal end.

Despite being described as common in many portions of their range, Great Plains Toads are fossorial and difficult to monitor except during spring breeding. In some portions of their range, populations may be scattered and isolated. Population densities are known to fluctuate widely in association with periodic droughts.

Great Plains Toads currently receive no federal or state protection in the U.S. They are, however, listed in Canada as a species of Special Concern under the Species at Risk Act. In Alberta, they are recognized as a species that may be at risk and in Manitoba as Threatened, but they are not listed as a species at risk in Saskatchewan.

ORIGINAL ACCOUNT by Brent M. Graves and James J. Krupa; photograph by Bruce Taubert.

Anaxyrus debilis Girard, 1854

Chihuahuan Green Toad

Chihuahuan Green Toads range from southwestern Kansas and adjacent southeastern Colorado south into Mexico (not shown) through central and western Texas, eastern and southern New Mexico, and southeastern Arizona.

Chihuahuan Green Toads are found in arid regions below elevations of about 1500 m in open, grassy plains, native prairie, desert grasslands, mesquite and creosote bush brushlands, and playa bottom grasslands, especially along the valleys of small creeks. Chihuahuan Green Toads often take refuge under rocks or in existing rodent or other burrows and may occur in grasslands that have been converted to agriculture but where herbicide and/or pesticide levels do not exceed lethal limits.

Localized populations of Chihuahuan Green Toads have likely declined in recent years because of the disappearance of their habitat, especially from the conversion and disappearance of wetlands and low-lying areas that the toads use in reproduction. Chihuahuan Green Toads are localized but frequently abundant in remaining areas of suitable habitat.

Chihuahuan Green Toads are listed as Protected in Kansas.

ORIGINAL ACCOUNT by Charles W. Painter; photograph by Mike Redmer.

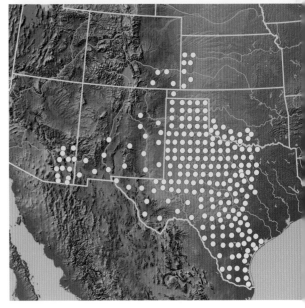

Chihuahuan Green Toad

Anaxyrus exsul Myers, 1942

Black Toad

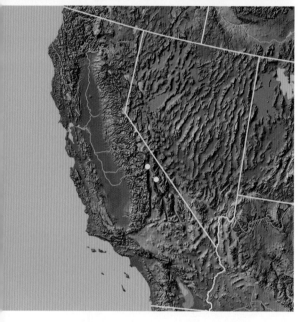

Black Toad

Black Toads are native only to the Deep Springs Valley in east central California. There is also a population of Black Toads inhabiting a flowing well near Salt Lake in Death Valley National Park that has been introduced.

Black toads inhabit the margins of marshes, streams, and sloughs associated with desert springs, where there is an abundance of wetland vegetation, including sedges, rushes, bulrushes, watercress, and rabbitsfoot grass. Because the surrounding desert is dry and cold, Black Toads are aquatic and may never go more than 12 m from standing water. Adults generally are diurnal but they may be active during the morning and early evening hours at air temperatures between 17 and 22°C.

The entire natural range of Black Toads encompasses approximately 15 hectares, one of the smallest ranges for any North American amphibian. As with any species with such a highly restricted distribution, there is concern that a catastrophic event or introduction of a disease or predator could eliminate the entire species.

Black Toads are listed as Threatened by the California Fish and Game Commission.

ORIGINAL ACCOUNT by Gary M. Fellers; photograph by Gary Nafis.

Anaxyrus fowleri Hinckley, 1882

Fowler's Toad

Fowler's Toads occur throughout most of eastern North America south of the Great Lakes, as well as the northern shore of Lake Erie, but largely excluding the southern Atlantic Coastal Plain and the Florida Peninsula. The western edge of the range of Fowler's Toads runs approximately from eastern Texas and Oklahoma northeast across central Missouri to southeastern Iowa, avoiding Kansas. This boundary, though, is ill defined because of the intergradation of Fowler's Toads with the more westerly occurring Woodhouse's Toads.

Fowler's Toads occur in areas with loose, well-drained gravelly or sandy soils, including sand dunes, sandy deciduous woodland, and rocky, poorly vegetated areas. They can be common along roadsides, near homes, and in fields, pastures, and gardens. Fowler's Toads are a typical species of the New Jersey Pine Barrens. They occur at elevations up to 1200 m in the Great Smoky Mountains of Tennessee but are much more common at lower elevations. Adult Fowler's Toads will return to their home ranges when displaced up to 1.28 km away and are known to occupy the same home range year after year.

Fowler's Toads may be extremely abundant throughout much of their range, especially in the northeast portion of their distribution. Populations vary widely in abundance in different years and at different places; at times there may be large numbers of individuals. However, Fowler's Toads have been extirpated from Nantucket and other, nearby, islands likely because of the over application of pesticides.

Fowler's Toad

Fowler's Toads are considered an Endangered Species in Canada under the Species at Risk Act and are likewise listed in Ontario under that province's Endangered Species Act.

ORIGINAL ACCOUNT by David M. Green; photograph by David M. Green.

Anaxyrus hemiophrys Cope, 1886

Canadian Toad

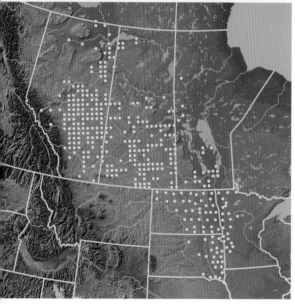

Canadian Toad

Canadian Toads are distributed from just south of Great Slave Lake in the Northwest Territories, across eastern Alberta and central Saskatchewan, into western and southern Manitoba, and south into the U.S. in eastern North and South Dakota and adjacent Minnesota, ranging from elevations of 300 to over 2100 m.

Canadian Toads occur in moderately moist prairie, aspen parkland, boreal forest, and pothole wetland country in places with enough water to keep such small wetlands from becoming saline. They can be found usually near streams, lakes, irrigation ditches, or flooded fields or, increasingly toward the north, in areas of grass or willow peatlands. During droughts, Canadian Toads may enter gutters and sewers in small towns. Canadian Toads are good swimmers and are more aquatic than most other toads.

Despite evidence for declines in Alberta, contractions and expansions of populations of Canadian Toads due to the availability of surface water are no doubt a natural occurrence. Droughts likely have a more profound effect on Canadian Toad populations in today's fragmented landscapes than they had in the past. Canadian Toad populations in the southern portion of their range experience periodic natural stress from severe droughts over a 10- to 11-year cycle.

Montana lists Canadian Toads as Endangered.

ORIGINAL ACCOUNT by Michael A. Ewert and Michael J. Lannoo; photograph by Constance Brown.

Anaxyrus houstonensis Sanders, 1953

Houston Toad

Houston Toads occur only on the coastal plain of southeastern Texas.

Houston Toads are restricted to areas with sandy, loose soils into which they can burrow. Many populations of Houston Toads are located in or near loblolly pine forests, likely because both Houston Toads and loblolly pines are both dependent on sandy soils.

Populations of Houston Toads have declined precipitously, and only a few scattered populations are known to still be extant. Population sizes appear to be small. The primary causes of the decline of Houston Toads have been prolonged droughts and various types of anthropogenic habitat modification, such as urbanization, recreational overdevelopment, road building, agriculture, and deforestation. Introduced red fire ants, which are known predators of recently metamorphosed toads, as well as increasing temperatures and aridity associated with climatic warming trends, may also threaten the future of the Houston Toad.

Houston Toads are listed as Endangered under the U.S. Endangered Species Act.

ORIGINAL ACCOUNT by Donald B. Shepard and Lauren E. Brown; photograph by Paul Crump.

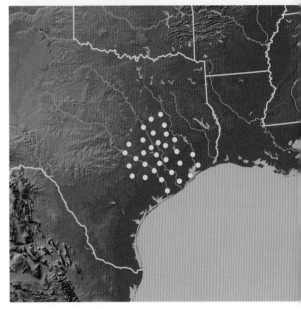

Houston Toad

Anaxyrus microscaphus Cope, 1867

Arizona Toad

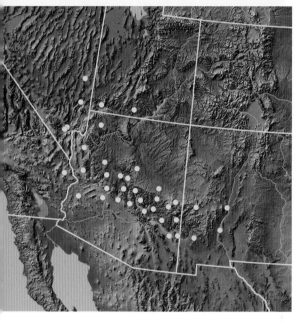

Arizona Toad

Arizona Toads range from the Mogollon Plateau of Arizona and southwestern New Mexico westward to the Colorado and Virgin River basins of northwestern Arizona, southern Nevada, southwestern Utah, and southern California. There are also populations in Mexico (not shown).

Arizona Toads inhabit sandy riparian areas characterized by dense willow clumps, open flats and flood channels, and terraces dominated by cottonwoods or live oaks, usually within 100 m or so of a stream. At elevations over 2000 m, though, Arizona Toads may move more widely into neighboring ponderosa pine forests during summer rains.

Arizona Toads appear to have declined in abundance at some sites following habitat disturbance because of damming, introduced predators, habitat degradation and golf courses along rivers and their tributaries. Damming along the Agua Fria River has reduced the extent of habitats favored by Arizona Toads for breeding.

Arizona Toads are listed as Endangered by the U.S. and as a Species of Special Concern by the state of California.

ORIGINAL ACCOUNT by Terry D. Schwaner and Brian K. Sullivan; photograph by Gary Nafis.

Anaxyrus nelsoni Stejneger, 1893

Amargosa Toad

Amargosa Toads are found only along a 16-km stretch of the Amargosa River and nearby spring systems in the Oasis Valley of southern Nevada.

Amargosa Toads require stretches of open, shallow water in slow-moving streams that persist long enough for their tadpoles to metamorphose into toadlets and emerge onto land. Adult toads forage at night along the water's edge and adjacent upland areas. During the day, Amargosa Toads typically find shelter in burrows, under debris piles, or amid dense vegetation. The Amargosa River channel and its riparian corridor serve as the primary route for the movements of Amargosa Toads between populations.

The greatest conservation concerns about Amargosa Toads are their limited distribution, their limited abundance, and the potential susceptibility of the toads' riparian habitat in a desert environment. Specific threats to Amargosa Toads include direct kill and habitat disturbance by wild burros, livestock, off-road vehicles, predation and competition by nonnative aquatic species, and alteration of riparian plant communities by invasion of tamarisk.

Amargosa Toads were at one time a candidate species for listing under the U.S. Endangered Species Act. They are currently classified in the U.S. as a Species of Concern and by the State of Nevada as a Protected species.

ORIGINAL ACCOUNT by Anna Goebel, Hobart M. Smith, Robert W. Murphy, and David J. Morafka; photograph by Gary Nafis.

Amargosa Toad

Anaxyrus punctatus Baird and Girard, 1852

Red-spotted Toad

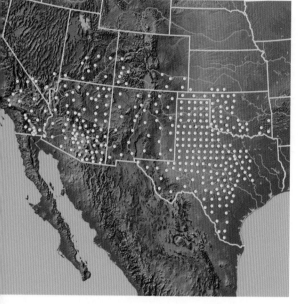

Red-spotted Toad

Red-spotted Toads are found from southern Kansas, Oklahoma, and central Texas, west to southeastern California. Their northern boundary includes southern Nevada, central Utah, and southern Colorado; their southern boundary is in northern Mexico (not shown). Red-spotted Toads occur at elevations from below sea level to almost 2000 m.

Adult Red-spotted Toads are typically associated with rocky streams in otherwise arid habitats, although they may occur away from streams in the well-drained, rocky soils of south central Arizona. They are tolerant of disturbed sites such as cattle tanks. In central Arizona and southern Nevada, Red-spotted Toads can be abundant in localized populations. Dispersal of toads among these populations appears to be frequent.

Widespread declines have not been noted for Red-spotted Toads, and they continue to be present at most historical localities, though they now appear absent from the vicinity of Austin, Texas.

Red-spotted Toads are listed by the State of Kansas as a species in need of conservation.

ORIGINAL ACCOUNT by Brian K. Sullivan; photograph by William Leonard.

Anaxyrus quercicus Holbrook, 1840

Oak Toad

Oak Toads are found on the southeastern Atlantic and Gulf Coastal Plain from the James River in southeast Virginia south and west to the Mississippi River in Louisiana. They are found throughout Florida and on some of the lower Florida Keys (not shown).

Oak Toads are most often associated with open canopied oak and pine forests that contain shallow, ephemeral wetlands including ponds, wet prairies, freshwater marshes, margins surrounding cypress swamps, and hardwood swamps. Oak Toads commonly seek refuge under boards and logs or in shallow depressions or burrows surrounded by vegetation, including cabbage palms and saw palmettos.

Oak Toads remain abundant in suitable, undisturbed habitats throughout their range. In Florida, development, drainage of wetlands, urbanization, and the introduction of exotic species via the pet trade always pose a potential threat and have substantially reduced natural areas where Oak Toads breed. Sensitive wetland habitats tend to be small and localized and are therefore more readily threatened by habitat destruction and fragmentation. Sedimentation and runoff of silt during highway and home construction projects, as well as pollution from agricultural pesticides and heavy metals, also pose serious threats to the quality of aquatic breeding sites and the survival of Oak Toad tadpoles.

Oak Toads are not listed either federally or at the state level in the U.S.

ORIGINAL ACCOUNT by Fred Punzo; photograph by Steve Bennett.

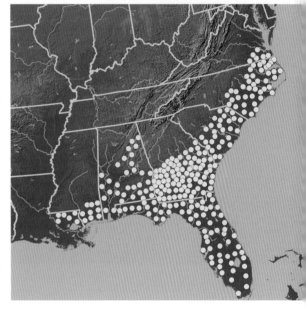

Oak Toad

Anaxyrus retiformis Sanders and Smith, 1951

Sonoran Green Toad

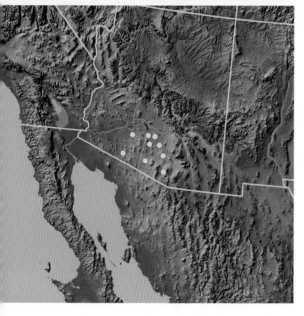

Sonoran Green Toad

Sonoran Green Toads are known from south central Arizona extending south into adjacent Mexico (not shown), at elevations from 150 to 900 m.

Adult Sonoran Green Toads have been observed in creosote flats, upland saguaro–palo verde associations, mesquite grasslands, and arid and semiarid grasslands. Large breeding aggregations of Sonoran Green Toads appear during and after summer monsoon storms in Arizona.

There do not appear to be any serious threats to populations of Sonoran Green Toads. The range of Sonoran Green Toads is limited to semiarid habitats and may be expanding because of irrigation associated with increasing agricultural activity.

Sonoran Green Toads are considered apparently secure but uncommon by the State of Arizona.

ORIGINAL ACCOUNT by Sean M. Blomquist; photograph by Bruce Taubert.

Anaxyrus speciosus Girard, 1854

Texas Toad

Texas Toads are distributed throughout central and western Texas, north into western Oklahoma, west into southeastern New Mexico, and south into Mexico (not shown).

Texas Toads are a desert species found in grassland, open woodland, and mesquite–savanna habitats in association with permanent streams, irrigation ditches, watering tanks, and ephemeral pools. Texas Toads prefer sandy soils, into which they burrow, as well as soils that are frequently flooded and have relatively high percentages of clay.

Population and abundance trends of Texas Toads throughout the majority of their range appear to be stable. They do not seem to be undergoing any dramatic declines, and they are considered to be one of most abundant toads in Texas. Nevertheless, declines in Texas Toad populations have been observed in the Rio Grande Valley because of the heavy use of pesticides and herbicides on agricultural lands.

Texas Toads are not listed at either the state or federal level in the U.S.

ORIGINAL ACCOUNT by Gage H. Dayton and Charles W. Painter; photograph by John Clare.

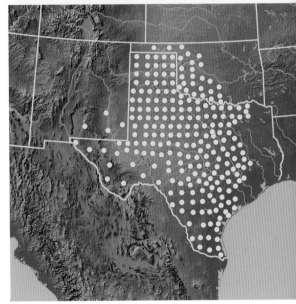

Texas Toad

Anaxyrus terrestris (Bonnaterre, 1789)

Southern Toad

Southern Toads occur primarily on the southeastern Coastal Plain from the James River in southeastern Virginia to the Florida Keys (not shown) and west through the Florida parishes of Louisiana to the Mississippi River. Southern Toads also occur outside of the Coastal Plain in portions of the Blue Ridge and Piedmont mountains of South Carolina and the Ridge and Valley region of Alabama.

Southern Toads are found in a variety of terrestrial habitats, including agricultural fields, pine woodlands, hammocks, and maritime forests. They prefer sandy soils, into which they burrow, or may take refuge under logs or other debris during the day. Southern Toads are common in backyards and gardens, tolerating humans well and, frequently, at night they may be found feeding on the insects attracted by lights.

Southern Toads are abundant throughout most of their range and are commonly seen, especially during the breeding season, despite massive habitat destruction in some areas. They have become increasingly rare in parts of Florida where nonnative Cane Toads have become established. The destruction of the preferred habitats of Southern Toads in southern Florida may have created better Cane Toad habitats, since direct competition between the two species is unlikely to be responsible for the observed population changes.

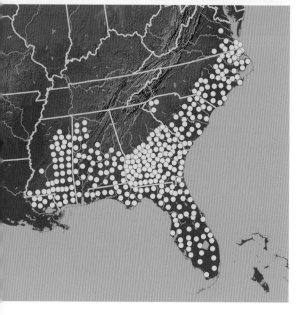

Southern Toad

Southern Toads are not listed in the U.S. under any state or federal regulations.

ORIGINAL ACCOUNT by John B. Jensen; photograph by Mike Redmer.

Anaxyrus woodhousii Girard, 1854

Woodhouse's Toad

Woodhouse's Toads occur throughout the Great Plains of the U.S. and southern Rocky Mountain region from Montana and North Dakota to the Gulf Coast of Louisiana, west to eastern California and southeastern Nevada, and south into central northern Mexico (not shown), with additional populations in western Idaho, southeastern Washington, and across central Missouri into Illinois. They range from near sea level at Salton Sea, California, to an elevation of almost 2500 m.

Particularly in the western parts of their range, Woodhouse's Toads are often associated with large river valleys at lower elevations, and moist meadows, ponds, lakes, and reservoirs at higher elevations. They can be common in disturbed habitats such as canals and irrigated fields, golf courses, and urban parks with ponds.

Widespread declines have not been noted for Woodhouse's Toads, though they do appear to now be absent from the Santa Cruz River floodplain in the vicinity of Tucson, Arizona, and from the vicinity of Austin, Texas. Local populations can be abundant, even in urbanized and other disturbed habitats. Woodhouse's Toads now occupy sites near Las Vegas that were not thought to be occupied historically, and also appear to be expanding their range in southeastern California in the vicinity of Palm Springs. In some degraded riparian areas, range expansions of Woodhouse's Toads are coincident with the decline of Arizona Toads. Similarly, Woodhouse's Toads appear to have expanded their range in agricultural areas formerly occupied by Red-spotted Toads in Colorado.

Woodhouse's Toads are listed as a Protected species in Oregon.

ORIGINAL ACCOUNT by Brian K. Sullivan; photograph by Gary Nafis.

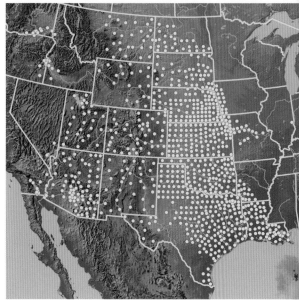

Woodhouse's Toad

Incilius alvarius Girard, 1859

Sonoran Desert Toad

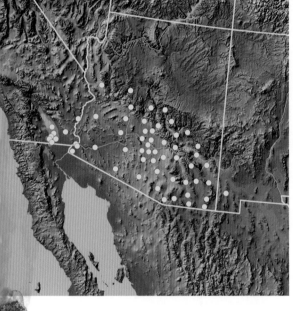

Sonoran Desert Toad

Sonoran Desert Toads are restricted to the Sonoran Desert region from extreme southwestern New Mexico throughout southern Arizona to southeastern California and into northwestern Mexico (not shown). Sonoran Desert Toads occur at elevations from near sea level to 1600 m.

Sonoran Desert Toads occur primarily in mesquite–creosote bush lowlands, but are also found in arid grasslands, rocky riparian zones with sycamore and cottonwoods, and oak–walnut woodlands in mountain canyons. During dry, pre-monsoon periods, Sonoran Desert Toads seek shelter, often in rodent burrows, rocky outcrops, or hollows under watering troughs, where they may remain for as long as 9 months. Prior to, or associated with, the onset of summer monsoon rains, though, adults move overland to breeding sites. The extent of these movements is not well known, although there are records of Sonoran Desert Toads moving over 400 m in a single day. Breeding usually occurs over 1 to 3 nights following a major rainfall in seasonal and permanent pools, irrigation ditches, and stock tanks.

Although Sonoran Desert Toads are abundant at many desert localities in Arizona, they appear to have declined in New Mexico and California.

Sonoran Desert Toads are listed as Endangered by the New Mexico Department of Fish and Game.

ORIGINAL ACCOUNT by M.J. Fouquette Jr., Charles W. Painter, and Priya Nanjappa; photograph by Bruce Taubert.

Incilius nebulifer Girard, 1854

Gulf Coast Toad

Gulf Coast Toads occur along the Gulf Coastal Plain of Texas, southern Louisiana, and extreme southwestern Mississippi and south into Mexico (not shown). They range westward along the Rio Grande and into the Hill Country of the Edwards Plateau region and Devil's River and Pecos River systems of Texas. Apparently isolated populations in southern Arkansas and northern Louisiana may represent dispersal along river systems.

Adult Gulf Coast Toads are frequently found in agricultural and wet hardwood areas, although they are relatively uncommon in pinelands. They are commonly seen at night in suburban areas foraging for prey on lawns and under streetlights. During the day, Gulf Coast Toads seek cover under logs; in rodent burrows, railroad ditches, roadside pools, garbage dumps, and storm sewers; under concrete slabs; or wedged into cracks and holes under sidewalks. Gulf Coast Toads have been observed 2 to 5 m above the ground in oak trees and may use tree holes repeatedly for periods of weeks. Adults may be found in vegetated areas of coastal barrier islands within 30 m of saltwater.

Gulf Coast Toads are frequently abundant, though local populations vary in size in relation to annual rainfall patterns. In some cases, their abundance may increase or decrease by as much as 50% from year to year. Gulf Coast Toads are adaptable and seem to tolerate habitat alterations caused by humans, including timber harvesting. They do not appear to be adversely affected by alteration of native vegetation or by invasion of nonnative vegetation.

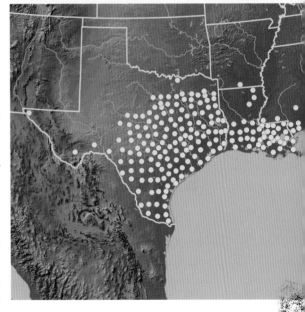

Gulf Coast Toad

Gulf Coast Toads are not listed in the U.S. under any state or federal regulations.

ORIGINAL ACCOUNT by Joseph R. Mendelson III; photograph by Brad Moon.

Rhinella marina Linnaeus, 1758

Cane Toad

Cane Toad

Cane Toads are native to an extensive area of South and Central America (not shown), extending north all the way to extreme southern Texas. Cane Toads were introduced in Puerto Rico and since the early 1930s they have been distributed throughout the island. They have also been introduced in southern Florida and throughout the Pacific Basin, including Hawaii. Cane Toads may occur from sea level to 1600 m. elevation.

Cane Toads are generally found along rivers and coasts in association with fresh and/or brackish water, including mangrove swamps. Adult Cane Toads survive in 10‰ seawater, but quickly die in 15‰ seawater. Cane Toads are tolerant of humans and are found in gardens, around houses, and in water tanks. They prefer disturbed areas and are rarely encountered in natural habitats. Cane Toads are nocturnal and are drawn to house and patio lights, which attract insects on which the toads feed. During the day, Cane Toads hide under rocks and boards, in burrows, and under long grass clumps out of direct sunlight.

Cane Toads do not usually appear in association with other species of frogs and toads. Either Cane Toads select bodies of water where other species are absent, or the other species are avoiding the Cane Toads. In the southern U.S., Cane Toads and Southern Toads are both present in some regions, but Southern Toads are found in drier pinelands and on drier ground within the Everglades than the areas frequented by Cane Toads. Cane Toads continue to be sold as pets, and releases and escapes facilitate their range expansion.

Cane Toads are not listed in the U.S. under any state or federal regulations.

ORIGINAL ACCOUNT by Jean-Marc Hero and Melody Stoneham; photograph by Mike Redmer.

Dendrobates auratus (Girard, 1855)

Green-and-black Poison Dart Frog

Native to Central America, Green-and-black Poison Dart Frogs were purposely introduced into Hawaii as part of a program to control non-native insects. The source of these frogs is thought be populations on the island of Taboga off the Pacific Coast of Panama. In Hawaii, Green-and-black Poison Dart Frogs currently reside only on Oahu in a few well-vegetated, moist valleys.

Green-and-black Poison Dart Frogs are found on the forest floor as well as in trees. They are diurnal, although in Hawaii they are less active on sunny afternoons.

Green-and-black Poison Dart Frogs are considered to be an introduced, invasive species on Oahu and are not listed in the U.S. under any state or federal regulations.

ORIGINAL ACCOUNT by Michael J. Lannoo and Priya Nanjappa; photograph by John Clare.

Green-and-black Poison Dart Frog

Craugastor augusti (Dugés, 1879)

Barking Frog

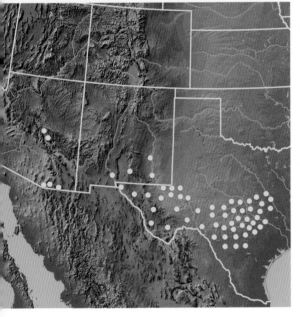

Barking Frog

Barking Frogs are found in central and south-eastern Arizona, south central and southeastern New Mexico, western and central Texas, and south into central Mexico (not shown), including the Santa Rita, Pajarito, Huachuca, Sierra Ancha, and Quinlan mountains, and along the Balcones Escarpment of Texas. They are found elevations between 900 and 1900 m.

Barking Frogs are terrestrial and commonly found in or near cliffs, caves, and limestone or other rock outcrops in Madrean evergreen woodland, creosote bush flats or juniper–oak scrub forest. They also have been found in caves and abandoned mines throughout their range.

In New Mexico and Texas, Barking Frog populations are mostly located on private lands, but they are also found on state, Fish and Wildlife Service, and National Park Service properties.

Barking Frogs are considered to be a Species of Special Concern in Arizona, which does not provide any legal protection.

ORIGINAL ACCOUNT by Cecil R. Schwalbe and Caren S. Goldberg; photograph by Caren Goldberg.

Eleutherodactylus coqui Thomas, 1966

Coquí

Coquís are native to Puerto Rico but have been unintentionally introduced into southeastern Florida and Hawaii, where they have rapidly expanded in range so that, now, they are found on the islands of Hawaii, Maui, Oahu, and Kauai.

In their native range of Puerto Rico, Coquís spend the day in the leaf litter on the ground and are arboreal at night. In Florida, they tend to forage up to 2.2 m off the ground and hide during the day in secluded areas such as rock piles, axils of bromeliad fronds, and tree holes or under rocks or logs.

In Puerto Rico, Coquís are widespread and abundant, and there is no evidence that they have declined in recent years except where habitats have been destroyed. Because Coquís are an introduced species in the U.S., concerns about their conservation have not been expressed. In Hawaii, where populations are expanding rapidly, instead the main concern is how to eradicate them.

ORIGINAL ACCOUNT by Margaret M. Stewart and Michael J. Lannoo; photograph by Alberto Puente-Rolon.

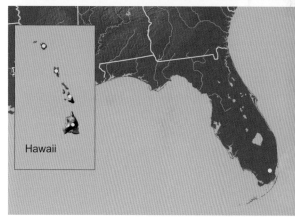

Hawaii

Coqui

Eleutherodactylus cystignathoides (Cope, 1878 "1877")

Rio Grande Chirping Frog

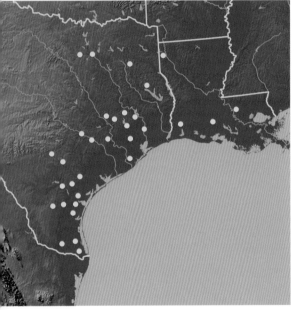

Rio Grande Chirping Frog

Rio Grande Chirping Frogs are native to southern Texas, where they occur along the lower Rio Grande Valley, and south into Mexico along the Gulf Coast (not shown). They have been introduced, probably by way of the potted plant trade, into numerous cities in southern and eastern Texas, as well as Louisiana, outside of their native range.

Rio Grande Chirping Frogs are able to acclimate to the urban settings into which they have been introduced, some very different from their native environment. They are often found in dense vegetation along streams and the edges of semipermanent ponds or pools, watered lawns, and gardens. Rio Grande Chirping Frogs can be found under debris piles or other cover during the day and are known to climb at night up to arboreal perches 0.2 to 1.5 m above the ground.

Rio Grande Chirping Frogs are abundant and thrive in the presence of humans.

Rio Grande Chirping Frogs have no federal or state conservation status in the U.S.

ORIGINAL ACCOUNT by J. Eric Wallace; photograph by Mike Redmer.

Eleutherodactylus guttilatus (Cope, 1879)

Spotted Chirping Frog

Spotted Chirping Frogs are known to occur in the Big Bend Region of western Texas, though the major part of their range lies to the south in Mexico (not shown).

In the mountains, Spotted Chirping Frogs inhabit rocky outcrops in ravines, along bluffs, and in man-made rock walls in oak–juniper woodland. At lower elevations, Spotted Chirping Frogs have been found in mines, along road cuts, and along limestone bluffs along the Rio Grande. Spotted Chirping Frogs seem to thrive with certain types of human-mediated habitat disturbance and may even prefer such artificial habitats as trail retaining walls and erosion-control barriers over nearby more natural habitats. Spotted Chirping Frogs are nocturnal and may be found by day under rocks, leaf litter, and debris.

Because of the remote nature of much of the native habitat for Spotted Chirping Frogs, they are most likely under no immediate population threats.

Spotted Chirping Frogs have no federal or state conservation status in the U.S.

ORIGINAL ACCOUNT by J. Eric Wallace; photograph by Tim Burkhardt.

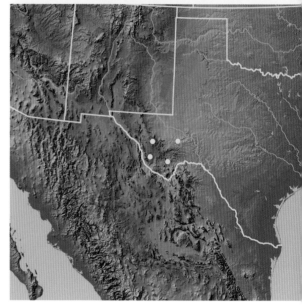

Spotted Chirping Frog

Eleutherodactylus marnockii (Cope, 1878)

Cliff Chirping Frog

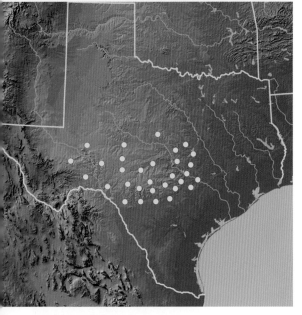

Cliff Chirping Frog

Cliff Chirping Frogs are found in central Texas on the Edwards Plateau west to the Stockton Plateau. They have also been accidentally introduced, through the potted plant trade, into some urban areas such as San Antonio.

Cliff Chirping Frogs are common in juniper- and oak-dominated landscapes, where they occur in association with rocky limestone ledges and bluffs, talus slides, ravines or streamsides, or in caves. They may be relatively abundant in urban areas, including city parks, suggesting an ability to cope with at least certain types of human-mediated habitat disturbance.

Cliff Chirping Frogs are active primarily at night during the hours after dusk, especially following spring and fall rains. During their breeding season, though, they may remain active throughout the night and males may continue to call during daylight hours. Occasionally, individuals have been observed on arboreal perches from 1.2 to 2.4 m above the ground. During the day, though, Cliff Chirping Frogs hide in rock crevices or under other cover objects.

Cliff Chirping Frogs have no federal or state conservation status in the U.S.

ORIGINAL ACCOUNT by J. Eric Wallace; photograph by William Leonard.

Eleutherodactylus planirostris (Cope, 1862)

Greenhouse Frog

Greenhouse Frogs are found naturally on Cuba and other islands in the West Indies as well as in southern Florida and Key West. Their present distribution in the continental U.S. also now includes populations all along the Gulf of Mexico coast from the Florida panhandle through southern Louisiana. They have been inadvertently introduced onto the islands of Hawaii and Oahu in Hawaii.

Greenhouse Frogs are found in a variety of humid terrestrial habitats and are particularly common in gardens, greenhouses, and nurseries. Greenhouse Frogs will hide during the day beneath leaf litter, mulch, boards, or stepping-stones. Greenhouse Frogs are also found outside of human influence in suitable natural habitats. They are secretive and are nocturnal except on warm, overcast, or rainy days.

Greenhouse Frog populations appear to be stable across much of their native range and have been rapidly expanding up peninsular Florida and along the Gulf Coast.

In Hawaii, Greenhouse Frogs are considered to be an undesired invasive species.

ORIGINAL ACCOUNT by Walter E. Meshaka Jr.; photograph by Mike Redmer.

Hawaii

Greenhouse Frog

Acris blanchardi Harper, 1947[2]

Blanchard's Cricket Frog

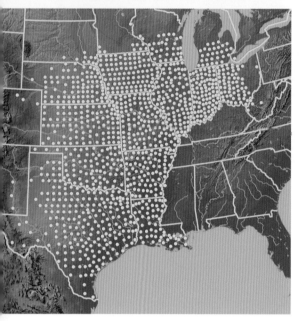

Blanchard's Cricket Frog

Blanchard's Cricket Frogs are distributed across the east central U.S. to the north and west of the Ohio and Mississippi River valleys from Ohio and western West Virginia west to Nebraska, south to western Texas and southeastern New Mexico, and into northern Coahuila, Mexico (not shown). Historically, they also occurred in Canada at Point Pelee and on Pelee Island in extreme southern Ontario.

Blanchard's Cricket Frogs are terrestrial and semiaquatic and may occur in or near freshwaters of any sort except large lakes and rivers.

Severe declines of Blanchard's Cricket Frogs throughout the northern extent of their range have resulted in scattered and isolated populations. A sharp decline in populations of Blanchard's Cricket Frogs has occurred throughout the upper Midwest of the U.S. Blanchard's Cricket Frogs declined in southern Ontario even in a National Park. This decline is alarming since, at the end of the 19th century and well into the mid-20th century, Blanchard's Cricket Frogs were evidently among the most numerous of all frogs in the midwestern U.S. There is no clear-cut indication of the cause(s) of this trend, although a number of anthropogenic factors and environmental conditions have been suggested. Blanchard's Cricket Frog populations remain stable in the more central regions of their range.

Blanchard's Cricket Frogs are listed as Endangered in Minnesota and Wisconsin and as a Species of Special Concern in Indiana, Michigan, and West Virginia. In Canada, they are listed as Endangered in Ontario and federally under the Species at Risk Act.

ACCOUNT by Robert H. Gray, Lauren E. Brown, and Laura Blackburn; photograph by Scott Gillingwater.

Acris crepitans Baird, 1854

Eastern Cricket Frog

Eastern Cricket Frogs are distributed in the southeastern U.S. east of the Mississippi River and south of the Ohio River to the edge of the Atlantic Coastal Plain and north up the Atlantic seaboard to southeastern New York. Isolated populations have been also reported on the Atlantic Coastal Plain in North Carolina and South Carolina.

Eastern Cricket Frogs can be found in or near virtually any body of freshwater, although they tend to be rare or absent at large lakes, wide rivers, or at polluted sites. Eastern Cricket Frogs typically occupy the damp zone along the water's edge, sitting on floating vegetation and debris, or on exposed banks. When approached by a potential predator, Eastern Cricket Frogs make several quick zigzag leaps, each often a meter or more in length, on the bank and/or floating vegetation. They then dive into the water and swim to the shoreline some distance away. After rains, Eastern Cricket Frogs disperse to more distant aquatic sites and often pass through drier, suboptimal terrain.

Eastern Cricket Frogs are listed as Threatened in New York.

ORIGINAL ACCOUNT by Robert H. Gray, Lauren E. Brown, and Laura Blackburn; photograph by Michael Graziano.

Eastern Cricket Frog

Acris gryllus (LeConte, 1825)

Southern Cricket Frog

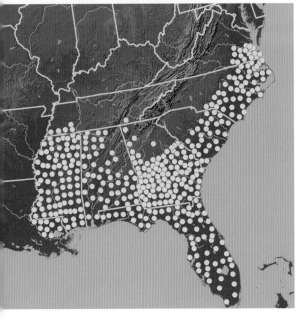

Southern Cricket Frog

Southern Cricket Frogs are found primarily on the Atlantic and Gulf Coastal Plains, from southeastern Virginia south and west to eastern Louisiana, including all of Florida and Mississippi. Populations are also known from southwestern and south central Tennessee, as well as in the Cumberland Plateau and Ridge and Valley region of Alabama and the Piedmont of Georgia.

Southern Cricket Frogs inhabit nearly every type of freshwater habitat, both temporary and permanent, with or without predatory fish, including roadside pools, ditches, swamps and marshes, and even interdunal pools within 18 m of the ocean. Southern Cricket Frogs are typically associated with more acidic waters than sympatric Eastern Cricket Frogs. Southern Cricket Frogs are occasionally observed in uplands distant from the nearest aquatic habitat, often in terrestrial meadows or wooded edges.

Southern Cricket Frogs were once considered to be the most abundant amphibian in the southeastern U.S., but there have been indications that Southern Cricket Frogs are declining in abundance locally. Breeding populations of Southern Cricket Frogs may occur at much lower densities on silviculture lands than in nearby native habitat because of elimination or severe alteration of the upland habitat resulting from intensive soil disturbance. Nevertheless, Southern Cricket Frogs remain common throughout most of their range.

Southern Cricket Frogs are not listed under any state or federal regulations in the U.S.

ORIGINAL ACCOUNT by John B. Jensen; photograph by Todd Pierson.

Hyla andersonii Baird, 1854

Pine Barrens Treefrog

Pine Barrens Treefrogs are known from the New Jersey Pine Barrens, the Fall Line sandhills between Moccasin Creek in North Carolina and the Congaree and Santee rivers in South Carolina, and the Florida Panhandle plus adjacent southern Alabama between the Escamba and Choctawatchee Rivers.

Pine Barrens Treefrogs inhabit shrub bogs dominated by mixtures of evergreen woody plants. They breed in bogs and swamps, where calling males and amplexed pairs are readily located.

Pine Barrens Treefrogs depend completely upon the long-term existence of their bog and swamp habitats.

Pine Barrens Treefrogs are currently considered to be Endangered by the State of New Jersey, Significantly Rare in North Carolina, Threatened in South Carolina and Alabama, and Rare in Florida.

ORIGINAL ACCOUNT by D. Bruce Means; photograph by Mike Redmer.

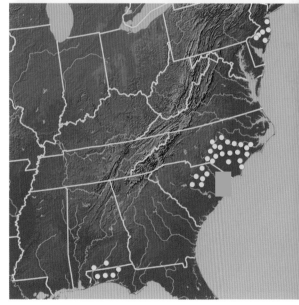

Pine Barrens Treefrog

Hyla arenicolor Cope, 1866

Canyon Treefrog

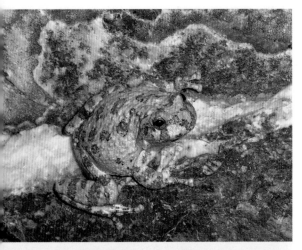

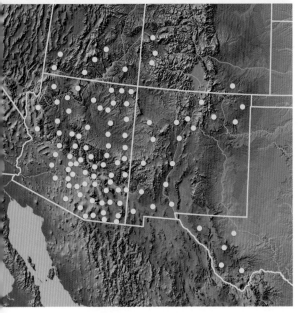

Canyon Treefrog

Canyon Treefrogs occur from southern Utah and southern Colorado, south through Arizona and New Mexico, and into northern Mexico (not shown) along the Sierra Madre Occidental. They also occur in southwestern Texas in the Chisos and Davis mountains west of the Pecos River.

Canyon Treefrogs are found along intermittent streams and permanent pools in deep, rocky canyons where there are cottonwood trees along the stream courses and pinyon–juniper woodlands along the slopes. They are often associated with areas of large boulders and rock outcrops. Canyon Treefrogs are active during warm rains but typically stay within a single leap of a pool. They are most active at night from May through December. Canyon Treefrogs are unusual in that they remain stationary throughout the day, with head down and limbs tucked under the body, on steep-sloped rocks in full sunlight within about 1 m from the water. In this resting position, they can maintain a body temperature between 29 and 31°C, which would be fatal to a typical, aquatic frog. An individual Canyon Treefrog will typically return to the same site over successive days.

Direct impacts from humans may constitute the biggest threats to Canyon Treefrogs, especially collectors and the increasing recreational use of public lands.

Canyon Treefrogs are listed as a Species of Special Concern in Colorado.

ORIGINAL ACCOUNT by Charles W. Painter; photograph by Bruce Christman.

Hyla avivoca Viosca, 1928

Bird-voiced Treefrog

Bird-voiced Treefrogs range from extreme south-western South Carolina, southwest across Georgia to the Florida Panhandle, west across the Gulf Coast to the east side of the Mississippi River drainage, and north through western Kentucky and Tennessee to extreme southern Illinois. Bird-voiced Treefrogs also occur west of the Mississippi River in isolated populations in central and north-western Louisiana, in the Red River Drainage of extreme southeastern Oklahoma, and in Arkansas.

Bird-voiced Treefrogs inhabit bottomland hardwood swamps and forested flood-plains, especially forests containing cypress and tupelo gum. Outside of the breeding season, adult Bird-voiced Treefrogs will perch on low vegetation, or reside on the ground under logs, in shrub thickets, or in tree crevices in or adjacent to the swamps they use for breeding.

The habitat specificity of Bird-voiced Tree-frogs may have had, and may continue to have, negative consequences for their survival. Populations may become isolated from one another because of drainage of swamplands.

Bird-voiced Treefrogs are currently listed as Threatened in Illinois,

ORIGINAL ACCOUNT by Michael Redmer; photograph by Mike Redmer.

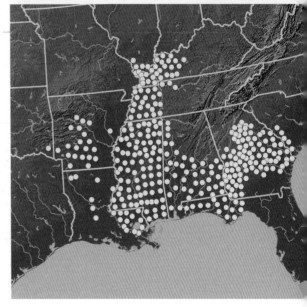

Bird-voiced Treefrog

Hyla chrysoscelis Cope, 1880

Cope's Gray Treefrog

Cope's Gray Treefrogs range throughout the southeastern U.S. from the Atlantic Coast approximately between the Delaware River and northern Florida's Withlacoochee River west to east central Texas and north into southern Manitoba, with populations also in Wisconsin, northern Illinois, and parts of Michigan.

Cope's Gray Treefrogs are found on trees or on mossy or lichen-covered fences, usually above ground, and hide during the day in knothole cavities and bluebird nesting boxes.

Cope's Gray Treefrogs are moderately tolerant of some pollutants, and they are tolerant of light habitat disturbance.

Cope's Gray Treefrogs are listed as Endangered in New Jersey.

ORIGINAL ACCOUNT by George R. Cline; photograph by Mike Redmer.

Hyla cinerea (Schneider, 1799)

Green Treefrog

Green Treefrogs range along the Atlantic Coast from Delaware Bay south through Florida, then west to the edge of the Edwards Plateau of Texas and the Red River drainage in southeast Oklahoma. They also are found in the Mississippi River drainage north to southeastern Missouri and southern Illinois. This species is expanding its range northward to include northeastern Oklahoma, central Missouri, southern and southwestern Illinois, and southern Indiana.

Green Treefrogs are typically found in habitats associated with permanent bodies of water containing abundant emergent vegetation, including swamps, sloughs, marshes, lakes, farm ponds, sewage ponds, fish-farm ponds, flooded borrow pits, flooded sinkholes, and ditches. They are commonly reported from barrier islands and other coastal areas, where they apparently are tolerant of brackish water. Green Treefrogs are frequently found around human dwellings and are known to winter in rock crevices, birdhouses, and among human litter such as tin cans.

Green Treefrogs are common or locally abundant throughout most of their range. Although they may be declining in one localized population in Florida, they have recently expanded their range in several bordering states.

Green Treefrogs are not listed under any U.S. state or federal laws.

ORIGINAL ACCOUNT by Michael Redmer and Ronald A. Brandon; photograph by Dante Fenolio.

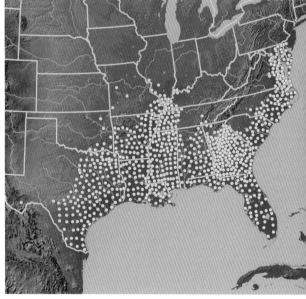

Green Treefrog

Hyla femoralis Bosc, 1800

Pine Woods Treefrog

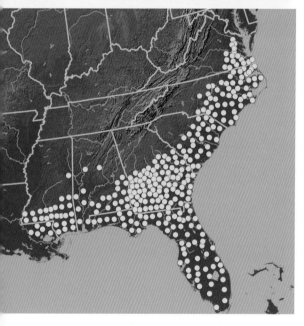

Pine Woods Treefrog

Pine Woods Treefrogs occur on the Coastal Plain from near the Mattaponi River in Virginia, south through most of Florida, including the Everglades, and west to the Mississippi River in Louisiana. There are also isolated, disjunct populations in central Alabama and Mississippi.

Pine Woods Treefrogs are strongly associated with pine forests, but also may be found in hammocks, swamps, cypress ponds, vernal pools, Carolina bays, mixed hardwood and pine forests, and brackish marshes. Pine Woods Treefrogs climb to the tops of longleaf pines, and seek shelter in cabbage palms, bromeliads, and pitcher plants.

Pine Woods Treefrogs are locally common across most of their range.

Pine Woods Treefrogs are not listed by any state in which they occur. However, Florida, Louisiana, Mississippi and Virginia regulate the number that can be collected and sold commercially.

ORIGINAL ACCOUNT by Joseph C. Mitchell; photograph by Brad Moon.

Hyla gratiosa LeConte, 1857

Barking Treefrog

Barking Treefrogs are found in the Coastal Plain of the eastern U.S. from Maryland, Delaware, and southern New Jersey south through most of the Florida Peninsula and westward to southeastern Louisiana. Inland, they range north thorough the Alabama Piedmont and western Tennessee to the western portion of the Mississippian plateau region in Kentucky.

Barking treefrogs breed in a wide variety of shallow wetlands, including ephemeral pools, semipermanent ponds, and permanent ponds. They usually remain in trees and shrubs or burrow into damp sand under logs or grass tussocks around the pond border when they are not engaged in calling or reproduction. They have been found up to 2.5 m above ground on vegetation or in trees. Adult Barking Treefrogs are not often encountered except after hard rains.

Barking Treefrogs are common in North Carolina, Florida, and Louisiana and uncommon to rare in Kentucky, Tennessee, Virginia, and other, more northern parts of their range. Habitat loss and the small number of occurrences put them at risk in the northern portions of their range.

Barking Treefrogs are listed as Endangered in Delaware and Maryland, Threatened in Virginia, and Legally Protected in Tennessee. The populations in Kentucky are not listed.

ORIGINAL ACCOUNT by Joseph C. Mitchell; photograph by Todd Pierson.

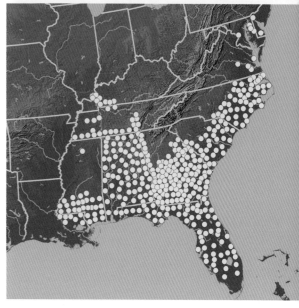

Barking Treefrog

Hyla squirella Bosc, 1800

Squirrel Treefrog

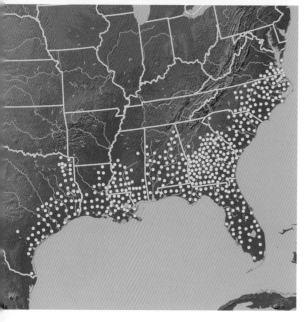

Squirrel Treefrog

Squirrel Treefrogs are found along the Atlantic Coastal Plain from Virginia to the Florida Keys (not shown) and along the Gulf Coastal Plain from south Florida to the Mexican border. They also occur on numerous barrier islands off the southeastern Atlantic Coast and Florida Gulf Coast.

Squirrel Treefrogs show little discrimination in their selection of habitat. They occur in and around buildings, in gardens, grasslands, weed or brush tangles, bottomland hardwoods, riparian zones, open woodlands, pinelands, trees, vines, cypress stands, longleaf pine–turkey oak–wiregrass associations, and longleaf pine–slash pine flatwoods. They even occupy coastal estuarine and harsh barrier island habitats. In other words, Squirrel Treefrogs occur almost any place associated with moisture, food, and cover, though they do exhibit some preference for open woodlands, such as mature pine and mixed hammock forests and open woody areas. They will seek shelter in trash piles and bromeliad plants, and often select narrow spaces in human structures. Squirrel Treefrogs will breed in rainwater pools, even those affected by saltwater spray in the Florida Keys. Squirrel Treefrogs are known for their ability to change colors to match their backgrounds.

Squirrel Treefrogs are among of the most common of frogs within their range. Populations in urbanized areas of southeastern Virginia have become extirpated, yet urban populations are relatively healthy in regions of Georgia and Florida.

Squirrel Treefrogs are not listed under any U.S. state or federal laws.

ORIGINAL ACCOUNT by Joseph C. Mitchell and Michael J. Lannoo; photograph by Mike Redmer.

Hyla versicolor LeConte, 1825

Gray Treefrog

Gray Treefrogs are found from east central Texas and adjacent Louisiana to the west and north of the Mississippi and Ohio River Valleys through eastern Oklahoma, western Arkansas, Missouri, and Illinois east to the Atlantic Coast between southern Virginia and New Brunswick. To the northwest, they extend to southwestern Manitoba and adjacent Ontario and, north of the Great Lakes, from about Sault Ste. Marie across southern Ontario to southwestern Québec.

Gray Treefrogs inhabit mixed hardwood and bottomland forests and woodlots, generally alongside rivers or swamps. They are usually found in trees or on mossy or lichen-covered fences, usually above ground. During the breeding season, male Gray Treefrogs call from emergent vegetation or from floating algae near the edges of ponds, ephemeral wetlands, and ditches. They may begin calling from high in the trees surrounding a pond and move progressively closer to the water's edge as the evening progresses. Gray Treefrog tadpoles inhibit the presence of tadpoles of other species of frogs.

Gray Treefrogs are the tetraploid counterparts to the near identical, but diploid, Cope's Gray Treefrogs (see page 54).

Gray Treefrogs are relatively tolerant of habitat disturbance by humans, appear to be moderately tolerant to an assortment of pollutants and are known tolerate pH levels as low as 3.5. Ultraviolet B radiation, though, is known to have a negative effect on the swimming activity of Gray Treefrog tadpoles, possibly because it can enhance the toxicity of the pesticide carbaryl.

Gray Treefrogs are not listed under any provincial, state, or federal regulations in either Canada or the U.S.

ORIGINAL ACCOUNT by George R. Cline; photograph by Scott Gillingwater.

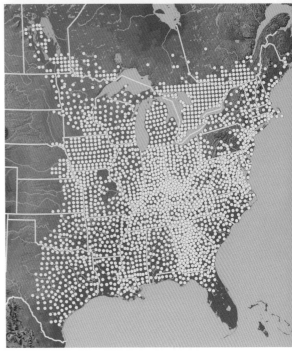

Gray Treefrog and Cope's Gray Treefrog

Hyla wrightorum Taylor, 1938

Arizona Treefrog

Arizona Treefrog

Arizona Treefrogs are found in the uplands of central Arizona from the Coconino Plateau south and east to the San Andres Mountains of west central New Mexico at elevations from 900 to 2900 m. Isolated populations occur in the Sierra Anchas Mountains in central Arizona and the Huachuca Mountains and adjacent Canelo Hills of southeastern Arizona. Arizona Treefrogs also occur in the Sierra Madre Occidental of northern Mexico, south to near Michoacán (not shown).

Arizona Treefrogs inhabit meadows or areas near slow-moving streams in pine–oak or pine–fir forests. Adults can be found in trees at some elevation off the ground. Survival of Arizona Treefrog tadpoles is greatly reduced in the presence of larval Western Tiger Salamanders.

Abundance may vary, but Arizona Treefrogs are common in appropriate habitats in west central New Mexico, but may be susceptible to extirpation in other parts of their range owing to small population sizes.

Arizona Treefrogs are not listed under any state laws. The U.S. Fish and Wildlife Service considers Arizona Treefrogs to be a candidate species under the Endangered Species Act.

ORIGINAL ACCOUNT by Erik W.A. Gergus, J. Eric Wallace, and Brian K. Sullivan; photograph by Bruce Taubert.

Osteopilus septentrionalis (Duméril and Bibron, 1841)

Cuban Treefrog

Cuban Treefrogs are native to Cuba, the Isle of Pines, the Cayman Islands, and the Bahamas, but they have been introduced into Puerto Rico, St. Croix, St. Thomas, Necker Island, and mainland Florida. The question of whether they are native to Key West is unresolved. On mainland Florida, Cuban Treefrogs rapidly dispersed from Miami to establish themselves as far north as the Florida Panhandle.

In the Everglades, Cuban Treefrogs occupy an otherwise underexploited niche with respect to habitat and diet, requiring only a vertical vegetative structure, water, refuges, and an invertebrate prey base to sustain populations. They are larger and more fecund than either Green Treefrogs or Squirrel Treefrogs, their likeliest potential anuran competitors. Both of these native species are eaten by large Cuban Treefrogs.

Cuban Treefrogs are considered an invasive species on mainland Florida. Their rapid expansion into a variety of both natural and disturbed habitats, and their potentially high densities, mean that Cuban Treefrogs have demonstrably negative impacts, both predatory and competitive, on native amphibians in Florida.

Cuban Treefrogs are not protected in the U.S.

ORIGINAL ACCOUNT by Walter E. Meshaka Jr.; photograph by Mike Redmer.

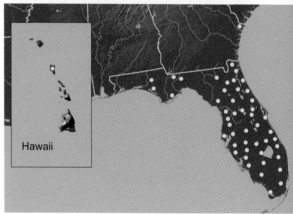

Hawaii

Cuban Treefrog

Pseudacris brachyphona (Cope, 1889)

Mountain Chorus Frog

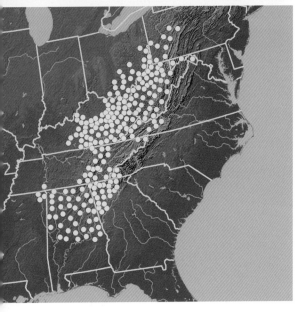

Mountain Chorus Frog

Mountain Chorus Frogs are distributed primarily from central western Pennsylvania southwestward through the Appalachians to northwestern Georgia, central Alabama, northeastern Mississippi, and adjacent Tennessee.

Adult Mountain Chorus Frogs move to foraging sites and shelters on hillsides in the forest after the breeding season.

Deforestation, urbanization, and loss of floodplain pools have apparently caused a decline in appropriate habitat for Mountain Chorus Frogs. Their historical range was undoubtedly wider than is currently known today, as there remain assorted disjunct populations in several states.

North Carolina lists Mountain Chorus Frogs as a Species of Special Concern and historical, given that it has not been seen in that state since the original observation. Mountain Chorus Frogs are not listed in any other state where they occur, but regulations in Mississippi, Pennsylvania, Tennessee, and Virginia restrict commercialization.

ORIGINAL ACCOUNT by Joseph C. Mitchell and Thomas K. Pauley; photograph by Michael Graziano.

Pseudacris brimleyi Brandt and Walker, 1933

Brimley's Chorus Frog

Brimley's Chorus Frogs are found along the Atlantic Coastal Plain from the Canoochee River in northeastern Georgia north through coastal Virginia to about the Rappahannock River.

Adult Brimley's Chorus Frogs are to be found well away from water in mixed pine and hardwood forests; pine forests; secondary dune scrub forest; forested wetlands dominated by red maple, loblolly pine, and sweetgum; and cultivated fields. They may be locally abundant.

Brimley's Chorus Frogs are not listed at either the state or federal level, though Virginia does not allow them to be exploited.

ORIGINAL ACCOUNT by Joseph C. Mitchell; photograph by Todd Pierson.

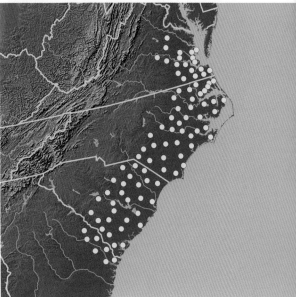

Brimley's Chorus Frog

Pseudacris cadaverina (Cope, 1866)

California Treefrog

California Treefrog

The distribution of California Treefrogs extends from the southern Sierra Lucia Range in coastal southern California through northern Baja California in Mexico (not shown). They occur across a wide range of elevations, from near sea level to around 2290 m.

Typical habitat for California Treefrogs includes clean rock surfaces, crevices, shade, and during the breeding season, quiet, clean water. They most commonly are found in close proximity to, and along, stream channels. During the daytime, individuals seek refuge in cavities or small depressions on the surfaces of the boulders lining streams, often fully exposed to direct sunlight. These perches are usually within a few jumps from the nearest pool. During the autumn and winter, California Treefrogs are occasionally found in upland habitats far from streams.

California Treefrogs have a discontinuous distribution within their range but are often locally abundant. Nevertheless, California Treefrog tadpoles are difficult to find in presumably high-quality habitat anywhere populations of non-native predatory fish have become established, suggesting that some populations may be experiencing declines.

California Treefrogs are not listed at the state or federal level.

ORIGINAL ACCOUNT by Edward L. Ervin; photograph by Bruce Christman.

Pseudacris clarkii (Baird, 1854)

Spotted Chorus Frog

Spotted Chorus Frogs are distributed from approximately the Saline River in central Kansas south through central Oklahoma, eastern New Mexico, and central Texas into extreme northeast Tamaulipas in Mexico (not shown).

Spotted Chorus Frogs are found principally in prairies and prairie islands in savannas. After emerging from hibernation, adult Spotted Chorus Frogs are thought to range through pastures and fields while foraging. They do not frequent stream margins or pools except to breed.

Little is known of the historical abundance of Spotted Chorus Frogs, though they may readily expand their range during dry years.

Spotted Chorus Frogs are not listed by any state or by the U.S. government. In Kansas, Spotted Chorus Frogs can be commercially collected for sale as fish bait.

ORIGINAL ACCOUNT by Michael J. Sredl; photograph by Paul Crump.

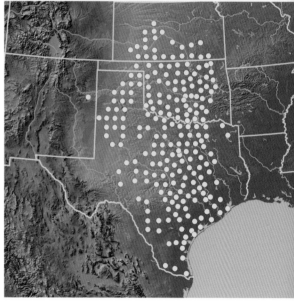

Spotted Chorus Frog

Pseudacris crucifer (Wied-Neuwid, 1838)

Spring Peeper

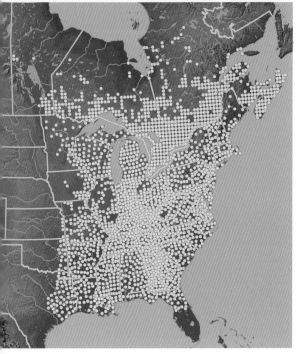

Spring Peeper

Spring Peepers are distributed throughout most of eastern North America east of a line from eastern Texas to the northern tip of Lake Winnipeg, except for the southern half of the Florida Peninsula. At their northern limit, Spring Peepers occur in Ontario and Québec up to the La Grande River on James Bay.

Spring Peepers generally are found in most eastern wooded habitats provided there are lowland marshes or other wetlands in which to breed. Toward the north, they are generally associated with sphagnum bogs, cattail wetlands, ponds, pools, and ditches in and near woods. In the south, though, Spring Peepers are found in low hammocks, swamp borders, the more open bay heads and tangles along smaller streams. During summer and fall, nonbreeding adults retreat to woodlands—where they are difficult to find—to feed and hide.

Spring Peepers are thought to be resistant to habitat fragmentation; however, populations of Spring Peepers have undoubtedly been lost because of logging, the conversion from forest to agricultural land use, mining activities, road building, urbanization, and suburbanization.

Spring Peepers are listed as Threatened in Kansas and Protected in New Jersey. These designations offer legal protection and require that permits be obtained before any activities involving Spring Peepers are undertaken in these states. In Canada, Spring Peepers are not listed under any provincial or federal laws.

ORIGINAL ACCOUNT by Brian P. Butterfield, Michael J. Lannoo, and Priya Nanjappa; photograph by Mike Redmer.

Pseudacris feriarum (Baird 1854)[3]

Upland Chorus Frog

Upland Chorus Frogs occur from southeastern Missouri, western Tennessee and eastern Louisiana east and north through Virginia, West Virginia and Maryland into central Pennsylvania. Their range includes all of the southeastern U.S. states but tends to exclude the Gulf and Atlantic Coastal Plains as well as the Appalachian highlands.

Upland Chorus Frogs are early spring breeders in seasonal and semipermanent fishless wetlands. During the late summer and early fall, single males or small numbers of males can frequently be heard calling from upland sites.

Although rarely seen outside the breeding season, Upland Chorus Frogs can be among the most abundant members of many amphibian assemblages. Spring calling in robust populations can be deafening, sometimes to the point of causing pain for nearby humans.

Upland Chorus Frogs are not listed under any state or federal laws.

ACCOUNT by David M. Green and Michael J. Lannoo; photograph by Mike Redmer.

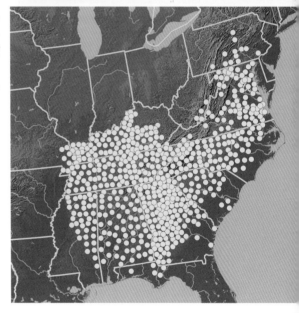

Upland Chorus Frog

Pseudacris fouquettei Lemmon, Lemmon, Collins and Canatella, 2008[4]

Cajun Chorus Frog

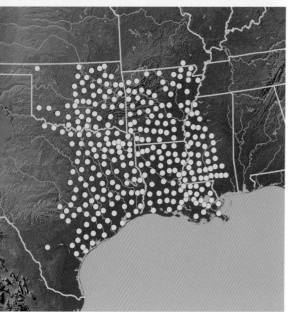

Cajun Chorus Frog

Cajun Chorus Frogs are found in western Mississippi, Louisiana, and Arkansas, eastern Texas and Oklahoma, and extreme southern Missouri.

As with their close relatives among the Chorus Frogs, Cajun Chorus Frogs are among the first spring breeders in any amphibian assemblage and, while rarely encountered, are usually among the most abundant of species. Adults are upland dwellers but do not travel great distances from breeding wetlands.

While there has been little call for conservation activities on behalf of Cajun Chorus Frogs, the small size and secretive nature of adults and postmetamorphic juveniles makes population assessments difficult outside of the breeding season.

Cajun Chorus Frogs are not listed under any U.S. state or federal laws.

ACCOUNT by David M. Green and Michael J. Lannoo; photograph by Michael Graziano.

Pseudacris hypochondriaca (Hallowell, 1854)

Baja California Treefrog

Baja California Treefrogs are found from southern Nevada and northwestern Arizona into coastal and southern California, except for desert regions. Their range includes the islands of Santa Rosa, Santa Cruz, and Santa Catalina. They also range into northern Baja California in Mexico (not shown). They are similar in appearance and closely related to Sierran Treefrogs, which are found immediately to the north.

Baja California Treefrogs inhabits grassland, chaparral, woodland, and desert oases, where they seek cover in rock fissures, under bark, in vegetation along streams, in rodent burrows, and in various other nooks and crannies. Baja California Treefrogs will breed in springs, ponds, irrigation canals, streams, and other small, semipermanent bodies of water.

Several localized populations of Baja California Treefrogs, such as those in California City, California, are thought to have been introduced through the plant trade. A single, obviously displaced Baja California Treefrog was found in the early 1980s in a plant shop in Estes Park, Colorado.

Baja California Treefrogs are not listed under either state or federal law.

ACCOUNT by David M. Green; photograph by Gary Nafis.

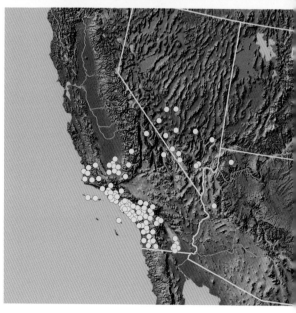

Baja California Treefrog

Pseudacris illinoensis Smith, 1951

Illinois Chorus Frog

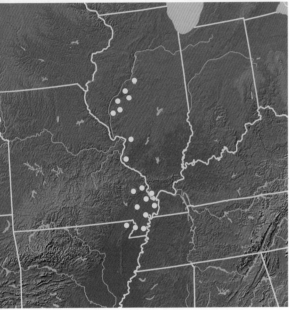

Illinois Chorus Frog

Illinois Chorus Frogs range from west central Illinois south along the Illinois and Mississippi river drainages into southeastern Missouri and northeastern Arkansas.

Illinois Chorus Frogs are found in close association with sand prairies, or sandy cultivated fields, flatwoods, and wooded floodplains. Much of their year is spent underground in burrows excavated using their forelimbs in areas devoid of, or lacking in, heavy vegetative cover. Populations of Illinois Chorus Frogs appear to be small, with choruses usually consisting of fewer than 20 males.

The primary cause of the decline of Illinois Chorus Frogs has been drainage and conversion of wetlands to agriculture. Many populations are apparently able to persist in disturbed habitats, but others continue to decline. Because most populations are small, they may be prone to local extirpations.

Illinois Chorus Frogs are listed as Threatened in Illinois, Rare in Missouri, and a Species of Special Concern in Arkansas. They have been designated as a Category 2 species by the U.S. Fish and Wildlife Service.

ORIGINAL ACCOUNT by Donald B. Shepard, Lauren E. Brown, and Brian P. Butterfield; photograph by Mike Redmer.

Pseudacris kalmi (Harper, 1955)[6]

New Jersey Chorus Frog

New Jersey Chorus Frogs are distributed throughout the Delmarva Peninsula in Virginia, in the
adjacent regions of Maryland and Delaware, and
on Staten Island, New York.

New Jersey Chorus Frogs are rarely encountered outside of the breeding season, and even
during breeding, calling males can be difficult
to locate.

New Jersey Chorus Frogs are not listed under
any U.S. state or federal laws.

ACCOUNT by David M. Green and Michael
J. Lannoo; photograph by Richard D. Bartlett.

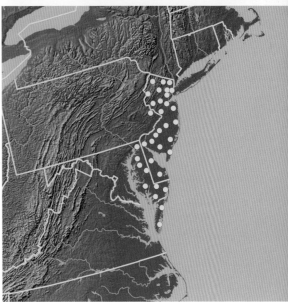

New Jersey Chorus Frog

Pseudacris maculata (Agassiz, 1850)[7]

Boreal Chorus Frog

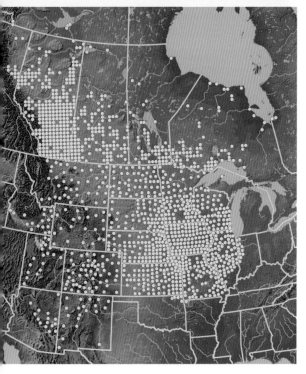

Boreal Chorus Frog

Boreal Chorus Frogs range from the McKenzie River Valley near to Great Bear Lake in the Northwest Territories, across to Hudson Bay in northern Manitoba and Ontario all the way to southern James Bay in Quebec, and southwest to Missouri, Kansas, and parts of New Mexico and Arizona. Their range lies to the east of the Rocky Mountains from northern British Columbia to Montana, but from Idaho south it extends west well into the mountains.

Boreal Chorus Frogs inhabit wet grassy meadows or wooded areas near ponds in prairie, parkland, boreal forest and foothills forest habitats, agriculture areas and cities. They can also be found on the tundra in the north of their range. Boreal Chorus Frogs are among the earliest of spring breeders, and will begin to call from almost any shallow, fishless, body of water, including roadside ditches, flooded fields and marshes, almost immediately after ice-off. Boreal Chorus Frogs and other, related species, produce antifreeze compounds that allow them to survive subfreezing conditions at or near the soil surface during winter.

Boreal Chorus Frogs are considered by Québec to be a species likely to be designated Threatened or Vulnerable.

ACCOUNT by David M. Green and Michael J. Lannoo; photograph by Randy Jennings.

Pseudacris nigrita (LeConte, 1825)

Southern Chorus Frog

Southern Chorus Frogs inhabit the Gulf Coastal Plain and Florida Peninsula from just north of the Tar and Pamlico rivers in North Carolina to the Pearl River in southern Mississippi, with isolated populations in Central Mississippi and southeast Virginia.

Southern Chorus Frogs are considered a pine savanna or pine flatwoods species. In Alabama, Southern Chorus Frogs are found where the soil is sandy and friable, whereas in southern Florida, Southern Chorus Frogs are always associated with limestone sinkholes, especially those bordering wet prairies, but are absent from wet prairies proper and from sandy country. Adults move to drier hammocks and ridges of the Pine Barrens outside of the breeding season.

There is no indication of any recent change in the distribution of Southern Chorus Frogs; however, their native longleaf pine flatwood habitat has been drastically reduced by conversion to slash pine plantations, by connection of ponds for drainage, and by fire-suppression measures. In the face of these activities, it is likely that Southern Chorus Frogs have experienced population reductions and extirpations.

Southern Chorus Frogs are not listed under either state or federal laws in the U.S.

ORIGINAL ACCOUNT by William T. Leja; photograph by Mike Redmer.

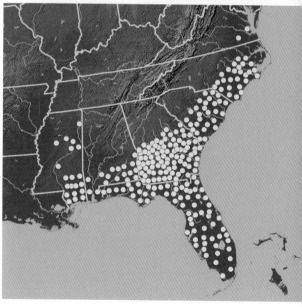

Southern Chorus Frog

Pseudacris ocularis (Bosc and Daudin, 1801)

Little Grass Frog

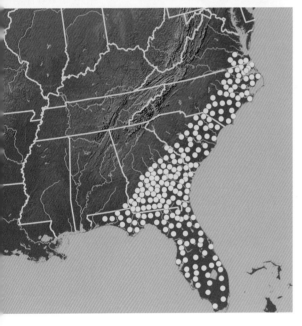

Little Grass Frog

Little Grass Frogs are found in the southeastern Coastal Plain of the U.S. from near the James River in southeastern Virginia to the southern tip of Florida and west to Choctawhatchee Bay in the Florida Panhandle. They have been reported from Key West, Florida; however, this record may be false. Misidentifications are not surprising, especially considering that the original description of Little Grass Frogs was based on a Cricket Frog.

Little Grass Frogs use grass, sedge, and/or sphagnum habitats in or near cypress ponds, bogs, pine flatwoods and savannas, river swamps, and ditches. They have been found calling from vegetation within brackish ditches, though it is unknown whether reproduction is attempted or successful in such waters. Adults are capable of climbing vines, tree trunks, and bushes to a height of 1.5 m or more. Little Grass Frogs are often active during both day and night.

Little Grass Frogs remain common throughout much of their range, and no substantial changes in their abundance have been noted.

Little Grass Frogs are not listed under either state or federal laws in the U.S.

ORIGINAL ACCOUNT by John B. Jensen; photograph by Joyce Marie Klaus.

Pseudacris ornata (Holbrook, 1836)

Ornate Chorus Frog

Ornate Chorus Frogs are restricted to the south-eastern Coastal Plain of the U.S. from extreme eastern Louisiana to North Carolina. The southernmost locality is in central Florida.

Ornate Chorus Frogs inhabit pine woodlands, pine–oak forests, and fallow fields. They need habitats with sandy substrates so that they can burrow; thus, the availability of sandy soils influences their distribution. Outside of the breeding season, Ornate Chorus Frogs are fossorial, often burrowed among the roots of herbaceous vegetation.

Ornate Chorus Frogs have been regarded as common throughout their range or even abundant. Recently, however, there have been indications that Ornate Chorus Frogs may be declining locally. Breeding populations of Ornate Chorus Frogs in the Munson Sand Hills of panhandle Florida occur in much lower densities in silvicultural lands than in nearby native habitat. It is likely that the accelerating conversion of natural pine habitats to industrial pine plantations throughout this species' range is reducing their abundance. Ornate Chorus Frogs are capable of reestablishing populations on abandoned agricultural lands.

Ornate Chorus Frogs are not listed under either state or federal regulations.

ORIGINAL ACCOUNT by John B. Jensen; photograph by Mike Redmer.

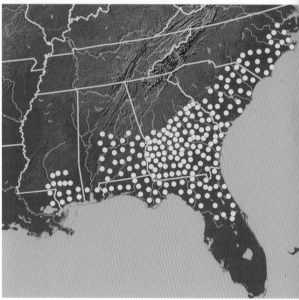

Ornate Chorus Frog

Pseudacris regilla Baird and Girard, 1852

Pacific Treefrog

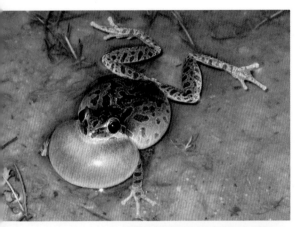

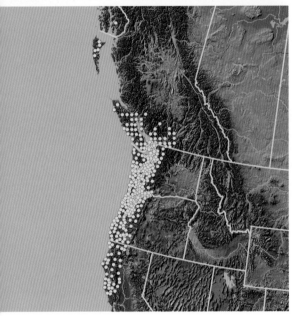

Pacific Treefrog

Pacific Treefrogs are found in southern British Columbia (including Vancouver Island), Washington, Oregon, and California. The map shows them as occurring west of the Cascades Range and through the Coast Range into northwestern California, although it is not yet established precisely where their range meets the range of Sierran Treefrogs to the east and south[5]. Pacific Treefrogs have been introduced to the Queen Charlotte Islands of British Columbia, and there are records from southeastern Alaska. Pacific Treefrogs occur at elevations from sea level to at least 1585 m in Washington.

Pacific Treefrogs exploit a great variety of habitats, including deserts, grasslands, mountains, and temperate rain forests. They use a variety of aquatic habitats for breeding and then move upland for the rest of the active season, where they may be found far from water. Pacific Treefrogs are found especially near streams, springs, ponds, wetlands, irrigation ditches, and other moist places. In these habitats, they hide among low plant growth, damp recesses among rocks and logs, under tree bark, in trees in damp forests, and in animal burrows in open country. They will move about in low shrubbery during moist weather but during dry periods or in dry habitats, Pacific Treefrogs tend to be more nocturnal, and seek moist, cool retreats for aestivation and for hibernation in the fall.

At most localities, Pacific Treefrogs are probably as common today as they were historically. Indeed, Pacific Treefrogs are usually the most abundant amphibians where they occur. With a few exceptions, the species is not declining and has not been targeted for conservation or recovery actions, though introduced goldfish will eat Pacific Treefrog eggs.

Pacific Treefrogs have no status under either the U.S. Endangered Species Act or the Canadian Species at Risk Act.

ORIGINAL ACCOUNT by James C. Rorabaugh and Michael J. Lannoo; photograph by Brad Moon.

Pseudacris sierra (James, Mackey, and Richmond, 1966)

Sierran Treefrog

Sierran Treefrogs are found from central California, Nevada, and western Utah north into eastern Oregon, Idaho, northwestern Montana and, probably, eastern Washington and British Columbia. The map depicts their distribution as being east of the Cascades Range in the U.S. and the Coast Mountains of British Columbia as far north as the Cariboo Country and Chilcotin, almost to the Fraser River Bend, though precisely where their range meets the range of Pacific Treefrogs to the west is not yet established[5]. In the Sierra Nevada of California, their distribution extends from sea level into mountains above timberline, indeed to all zones below Alpine–Arctic regions. Sierran Treefrogs may have been introduced in Arizona via plant nurseries near Phoenix and to livestock waters in the Virgin Mountains. Some populations in southern California may also be introduced.

Populations of Sierran Treefrogs are found associated with grasslands, forests, old fields, and agricultural areas. In high-elevation lakes of the Sierra Nevada, Sierran Treefrog tadpoles are not found in lakes that support salmonid fishes such as Rainbow Trout or Brook Char.

Where they occur, Sierran Treefrogs are typically one of the most common amphibians, often exhibiting robust populations. With a few exceptions, the species is not declining and has not been targeted for conservation or recovery actions.

Sierran Treefrogs have no status under the U.S. Endangered Species Act.

ACCOUNT by David M. Green; photograph by Gary Nafis.

Sierran Treefrog

Pseudacris streckeri Wright and Wright, 1933

Strecker's Chorus Frog

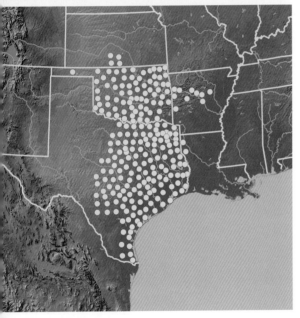

Strecker's Chorus Frog

Strecker's Chorus Frogs are found from extreme south central Kansas, south through Oklahoma, northwestern Louisiana, and Texas to the Gulf of Mexico, and east through central Arkansas.

Adult Strecker's Chorus Frogs are found in or near sand prairies or in sandy cultivated fields, flatwoods, and wooded floodplains. Strecker's Chorus Frogs spend much of their year underground in burrows they excavate using their forelimbs. Burrows tend to be in areas lacking heavy vegetative cover. Many populations of Strecker's Chorus Frogs are apparently able to persist in disturbed habitats, though others are in decline.

Strecker's Chorus Frogs are listed as Threatened in Kansas, a species of Special Concern in Arkansas and Louisiana, and Rare in Missouri.

ORIGINAL ACCOUNT by Donald B. Shepard, Lauren E. Brown, and Brian P. Butterfield; photograph by Mike Redmer.

Pseudacris triseriata (Wied-Neuwied, 1838)

Western Chorus Frog

Western Chorus Frogs range from southern Illinois, western Kentucky, and north central Tennessee northeast into most of Indiana, all of the Lower Peninsula of Michigan and Ohio, western Pennsylvania and New York, and southern Ontario east to southern Québec, northern New York, and northwestern Vermont[8]. Western Chorus Frogs are limited to the east by the Appalachian Mountains.

Western Chorus Frogs occur in a variety of habitats, including flood plains and humid forest. Outside of the breeding season, Western Chorus Frogs may be found hidden in leaf litter, among dead vegetation, in cracks in the ground, under logs, in crayfish burrows, or under woody debris.

Although Western Chorus Frogs appear to be tolerant of human activities, they are susceptible to agricultural chemicals and to baitfish and gamefish introductions into breeding wetlands. As they breed in small, insignificant, and ephemeral wetlands, they are vulnerable to habitat destruction due to urban and suburban development.

The eastern populations of Western Chorus Frogs from Lake Huron in Ontario to the St. Lawrence Valley in Québec are considered Threatened under the Canadian Species at Risk Act. Provincially, they are listed by Québec as *"vulnérable."*

ORIGINAL ACCOUNT by Emily Moriarty and Michael J. Lannoo; photograph by Mike Redmer.

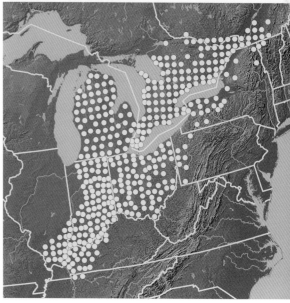

Western Chorus Frog

Smilisca baudinii (Duméril and Bibron, 1841)

Mexican Treefrog

Mexican Treefrog

In the U.S., scattered populations of Mexican Treefrogs are found in southern Texas, which represents the northern extreme of a distribution that extends south to Costa Rica (not shown). Mexican Treefrogs inhabit lowlands and foothills below elevations of 2000 m, but most localities are below 1000 m.

In the arid and semiarid places where they are found, Mexican Treefrogs inhabit forested and brushy areas around streams, resacas, and roadside ditches. Individuals have been observed living in the tops of palm trees. Mexican Treefrogs are nocturnal and are most active following rains. They seek shelter from hot and dry conditions under loose tree bark, in tree holes, in damp soil, and in the leaves of banana plants, bromeliads, and heliconias.

Mexican Treefrogs are a widely distributed species in Central America, where they are considered to be a common species, but populations in Texas are thought to be small and their distribution patchy.

Mexican Treefrogs are listed as Threatened by the State of Texas, but no federal protection exists or has been proposed.

ORIGINAL ACCOUNT by John H. Malone; photograph by Gary Nafis.

Smilisca fodiens Boulenger, 1882

Lowland Burrowing Treefrog

Lowland Burrowing Treefrogs are found from southern Arizona southward through western Sonora to Michoacán in Mexico (not shown), at sites ranging in elevation from sea level to 1500 m.

In the U.S., Lowland Burrowing Treefrogs are known to occur only in washes associated with river valleys, and little is known of their habitat preferences and ecology. Their ability to excavate shallow burrows and form an epidermal cocoon reduces evaporative water loss during long dry periods. Because these burrows are relatively shallow, Lowland Burrowing Treefrogs may be restricted to subtropical and tropical habitats, where soil temperatures do not drop below freezing.

Lowland Burrowing Treefrogs seem to be present at most known historical localities, suggesting that widespread declines have not occurred, but no studies of direct counts of breeding aggregations or population estimates have been conducted. Large breeding aggregations have been observed at many Arizona localities over the past 30 years. However, their U.S. distribution is restricted, making them vulnerable.

Lowland Burrowing Treefrogs are not offered legal protection in the U.S.

ORIGINAL ACCOUNT by Michael J. Sredl; photograph by Bruce Taubert.

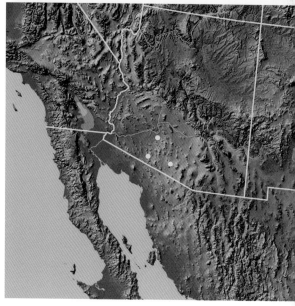

Lowland Burrowing Treefrog

Leptodactylus fragilis (Brocchi, 1877)

Mexican White-lipped Frog

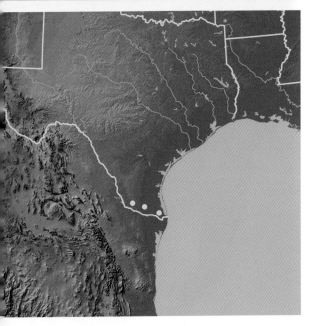

Mexican White-lipped Frog

Mexican White-lipped Frogs are known in the U.S. only in southernmost Texas in the extreme southern edge of the Lower Rio Grande Valley, but they also occur to the south throughout lowland Middle America to the north coast of South America as far as Venezuela (not shown).

Mexican White-lipped Frogs have been found in a variety of habitats wherever there is sufficient moisture, including semipermanent water bodies such as prairie potholes, oxbow lakes, and resacas. Mexican White-lipped Frogs may also be encountered in irrigated agricultural fields, irrigation ditches, low grasslands, and runoff areas. They are nocturnal and hide in burrows during the day.

Mexican White-lipped Frogs may now be extirpated in the Rio Grande Valley because of the continuous dispersal of organophosphate chemicals.

Mexican White-lipped Frogs are state-listed as Threatened, and are therefore protected, by the State of Texas.

ORIGINAL ACCOUNT by W. Ronald Heyer; photograph by Esteban Alzate.

Gastrophryne carolinensis Holbrook, 1836

Eastern Narrow-mouthed Toad

Eastern Narrow-mouthed Toads occur throughout the southeastern and lower midwestern U.S., from Maryland south to the Florida Keys (not shown), and west to central Texas and eastern Oklahoma, though they are absent from most of the Blue Ridge Mountains and the Appalachian region north of Tennessee. Eastern Narrow-mouthed Toads are found on numerous barrier islands in the Gulf Coast and off the southeastern Atlantic Coast (not shown). They occur as elevations as high as 730 m.

Eastern Narrow-mouthed Toads occur in a variety of habitats, including cypress gum swamps, live-oak ridges, pine oak uplands, sandy woodlands, prairies, mixed hardwood forests, riparian floodplains, brackish marshes, coastal scrub forest, and maritime forests. They tend to remain in the vicinity of breeding pools awaiting high humidity and heavy rains. Eastern Narrow-mouthed Toads are solitary, secretive, and highly terrestrial. Rocks, decaying logs, mats of vegetation, bark of logs and stumps, crayfish burrows, and boards along the edges of ponds and streams or around human dwellings are often used for shelter. Adults are tolerant of brackish water, which has allowed them to colonize barrier islands and live in brackish marsh habitats.

Forestry operations and urbanization can affect abundance of Eastern Narrow-mouthed Toads. In southern Maryland, ditching, agriculture, and logging are responsible for the reduction of suitable habitat. A disjunct population of Eastern Narrow-mouthed Toads in southeastern Iowa may be extirpated.

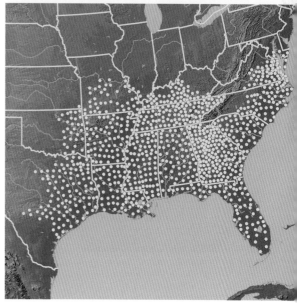

Eastern Narrow-mouthed Toad

Eastern Narrow-mouthed Toads are listed as Endangered by the state of Maryland.

ORIGINAL ACCOUNT by Joseph C. Mitchell and Michael J. Lannoo; photograph by Mike Redmer.

Gastrophryne olivacea (Hallowell, 1857 "1856")

Western Narrow-mouthed Toad

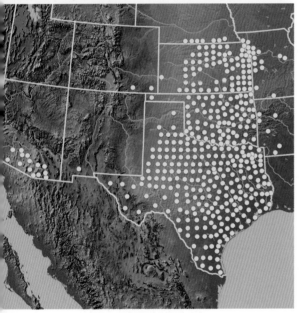

Western Narrow-mouthed Toad

Western Narrow-mouthed Toads are found from central Missouri along the Missouri River valley and extreme southern Nebraska; south through most of Kansas, Oklahoma, and Texas; and south onto the Mexican Plateau (not shown). Disjunct populations occur in the Oklahoma Panhandle, southeastern Colorado, western Kansas, central Arkansas, southwestern New Mexico, and southern Arizona. Western Narrow-mouthed Toads range from elevations of sea level to 1250 m.

Western Narrow-mouthed Toads are found in a variety of habitats, including prairies, desert grasslands, open woodlands, and oak woodlands. They are terrestrial, nocturnal, and secretive and usually occupy burrows and retreats under rocks, tree bark, roots, logs and leaf litter, or in mud cracks, usually near the vicinity of water.

The range of Western Narrow-mouthed Toads appears to have expanded recently into central Kansas along the Missouri River, and it is possible that as more land in Texas is converted to agriculture, the range of Western Narrow-mouthed Toads could increase there as well.

Western Narrow-mouthed Toads are included on the Arizona Department of Game and Fish's list of Threatened Native Wildlife as a Candidate species. They are also listed as a Species of Special Concern in Colorado and are given Endangered status in New Mexico.

ORIGINAL ACCOUNT by Michael J. Sredl and Kimberleigh J. Field; photograph by Bruce Taubert.

Hypopachus variolosus (Cope, 1866)

Sheep Frog

Sheep Frogs are at the northern limit of their distribution in southern Texas, though they range southward through Mexico to Costa Rica (not shown). Sheep Frogs are limited to regions of high relative humidity and therefore do not extend inland from the Gulf of Mexico more than 160 km.

Sheep Frogs occur in warm temperate and tropical thorn scrub and savanna habitats and are most frequently encountered in open woodlands or pasturelands with abundant shortgrass cover. In these habitats, they hide in moist subterranean burrows. Sheep Frogs are also commonly found in vegetative debris near ponds and irrigation ditches. Though they are commonly found after sufficient rainfall in areas with intact native brushland, Sheep Frogs have never been considered abundant in the U.S.

Sheep Frogs are currently listed as Threatened by the Texas Parks and Wildlife Department and are therefore protected from collection. They are not listed by the U.S. Fish and Wildlife Service, nor are they proposed for listing.

ORIGINAL ACCOUNT by Frank W. Judd and Kelly J. Irwin; photograph by Gary Nafis.

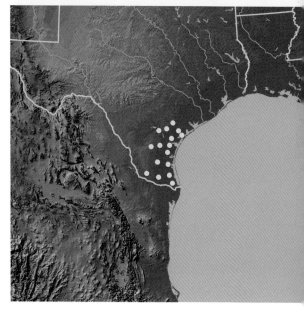

Sheep Frog

Scaphiopus couchii Baird, 1854

Couch's Spadefoot

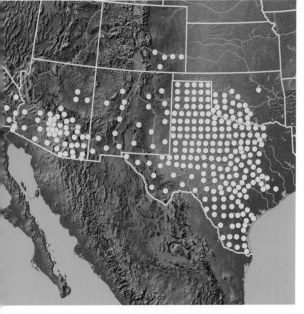

Couch's Spadefoot

Couch's Spadefoots range from central Texas and southwestern Oklahoma, west through central New Mexico and Arizona into southeastern California, and south into Mexico, including much of Baja California (not shown). Isolated populations occur in southeastern Colorado.

Couch's Spadefoots occur in mesquite and mesquite–yucca habitats, shortgrass plains, and creosote desert, as long as temporary rain-filled pools exist. The presence of sandy, well-drained soils is important to them. During the period of summer showers, Couch's Spadefoots reside in shallow soil-filled summer burrows 1.3 to 10 cm deep, often under dense vegetation, and surface activity is restricted to short periods following rains. Couch's Spadefoots spend 8 to10 months of the rest of the year 20 to 90 cm deep in soil-filled winter burrows, which they dig themselves. Thus, most of their life is spent in underground retreats.

Couch's Spadefoots have been eliminated wherever urban development and irrigated agriculture have destroyed their habitats. However, they will readily breed in ephemeral, artificial impoundments such as stock tanks and pools that may form beside roads and railroad grades. This has enabled them to colonize many areas where natural pools are rare or nonexistent. Their current distribution therefore differs somewhat from the past in that it reflects the effects both of habitat destruction and the spadefoots' colonization of new areas.

Couch's Spadefoots are listed as a Species of Special Concern in California and Colorado.

ORIGINAL ACCOUNT by Steven R. Morey; photograph by Mike Redmer.

Scaphiopus holbrookii Harlan, 1935

Eastern Spadefoot

Eastern Spadefoots range from Massachusetts and southeastern New York, south through the Atlantic Coastal Plain to the Florida Keys (not shown), and west to southeastern Louisiana, southeastern and northeastern Arkansas, and southeastern Missouri. They range northward through western and eastern Kentucky into southwestern Illinois, southern Indiana, southeastern Ohio, and northwestern West Virginia. Because of their cryptic habits and brief, irregular breeding bouts, the presence of Eastern Spadefoots is difficult to ascertain, and the peripheral extent of Eastern Spadefoots is still being determined.

Eastern Spadefoots occur in open and forested uplands and bottomlands, including disturbed habitats that have friable, sandy to loamy soils. Individual Eastern Spadefoots can sometimes be found at the surface under logs, and they use the same burrow for as long as 5 years.

In many places, Eastern Spadefoots no longer inhabit historical portions of their range because of losses of habitat.

Eastern Spadefoots are considered Rare in Missouri, a Species of Special Concern in Indiana, and Endangered in Ohio.

ORIGINAL ACCOUNT by John G. Palis; photograph by Mike Redmer.

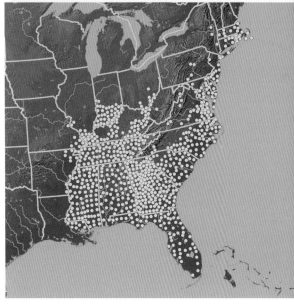

Eastern Spadefoot

Scaphiopus hurterii Strecker, 1910

Hurter's Spadefoot

Hurter's Spadefoots range from central Louisiana west to the Balcones Escarpment of the Edwards Plateau in Texas, and from eastern Oklahoma and north central Arkansas south to the Rio Grande.

Most aspects of the ecology of Hurter's Spadefoots are similar to those of Eastern Spadefoots. However, unlike Eastern Spadefoots, Hurter's Spadefoot tadpoles readily eat live invertebrate prey, including mosquito larvae, and fairy shrimp.

There are no state or federal regulations in the U.S. that apply to Hurter's Spadefoots.

ORIGINAL ACCOUNT by John G. Palis; photograph by Paul Crump.

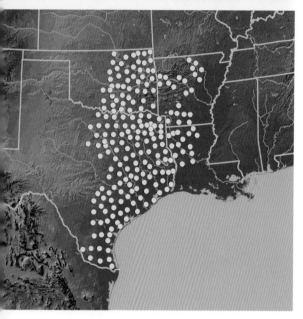

Hurter's Spadefoot

Spea bombifrons Cope, 1863

Plains Spadefoot

Plains Spadefoots range from southern Alberta, Saskatchewan, and southwestern Manitoba to northern Mexico (not shown), west into eastern Arizona, east to Nebraska and into the Loess Hills in western Iowa, Oklahoma, and along the Missouri River floodplain to its confluence with the Mississippi, and into central and southern Texas. Disjunct populations occur in extreme southern Texas and northeast Mexico, along the Arkansas River in Arkansas, and in southern Colorado.

Plains Spadefoots require loose, well-drained soils such as those found in floodplains, prairies, or loess hills in the northeastern part of their range, and grasslands, sandhills, semidesert shrub, and desert scrub in the southwest. Newly metamorphosed Plains Spadefoots burrow in mud along the edge of their natal ponds, hide in cracks in the hard earth, or seek cover in litter near the breeding site. Adult Plains Spadefoots are nocturnal and feed on the surface under humid conditions. When not feeding, they occupy shallow summer burrows or deeper winter burrows.

Plains Spadefoots tend to be locally abundant, but their localized distribution and secretive nature make them difficult to count. In parts of their range, natural breeding habitat has been severely reduced by agriculture, industrial, and other types of floodplain development.

Plains Spadefoots are listed as Protected in Manitoba but are federally listed in Canada as Not at Risk. They currently are not listed by any state or by the U.S. government.

ORIGINAL ACCOUNT by Eugenia Farrar and Jane Hey; photograph by Mike Redmer.

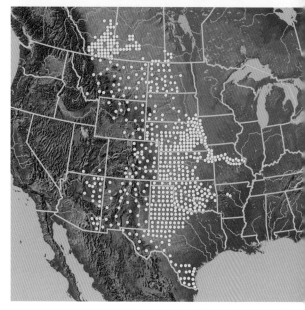

Plains Spadefoot

Spea hammondii (Baird, 1859, "1857")

Western Spadefoot

Western Spadefoots are found in the Great Central Valley of California and associated foothills, as well as the Coast Range east and south of San Francisco Bay. They also occur in northwestern Baja California in Mexico (not shown). Apparent gaps in the distribution are probably artifacts of uneven survey efforts, particularly in the northern portion of their range. Western Spadefoots usually are found at elevations below 365 m, but they have been observed as high as 1365 m in the mountains of southwestern California.

Western Spadefoots inhabit grasslands, oak woodlands, and occasionally coastal sage scrub or even chaparral in the vicinity of pools suitable for breeding. For several months following the first rains of autumn, Western Spadefoots spend periods of inactivity in shallow winter burrows. If it is not too cold or too dry, individuals can be encountered just after sunset at their burrow entrance with only their eyes protruding from the soil. Western Spadefoots also diminish their surface activity during the unbroken hot, dry periods of late spring, summer, and fall and lie quiescent in earth-filled burrows they construct themselves. Western Spadefoots may burrow over a meter deep and survive periods of osmotic stress during long periods of dormancy by accumulating urea in their body fluids.

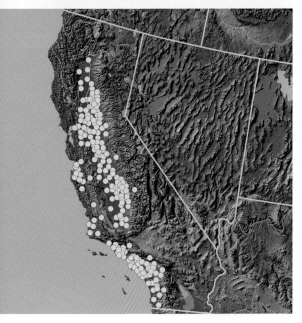

Western Spadefoot

Western Spadefoots have been eliminated from some parts of their range by urban and agricultural development. The prospects for Western Spadefoots seem to be dependent primarily on the balance between such developed areas and the undeveloped places where they can live.

Western Spadefoots are listed as a Species of Special Concern in California and federally in the U.S. as a Species of Concern.

ORIGINAL ACCOUNT by Steven R. Morey; photograph by Dante Fenolio.

Spea intermontana (Cope, 1883)

Great Basin Spadefoot

Great Basin Spadefoots have a broad distribution from south central British Columbia south into eastern Washington, Oregon, and California through Nevada, northwestern Arizona and Utah, into southern Idaho, northwestern Colorado, and southwestern Wyoming.

Great Basin Spadefoots are found primarily in sagebrush country, but they also may inhabit bunchgrass prairie, alkali flats, semidesert shrublands, pinyon juniper woodland, open ponderosa pine rangelands, and even high elevation spruce fir forests. They breed in springs, sluggish streams, and other permanent or ephemeral water sources. In the Bonneville Basin of Utah, over half of the breeding sites are manmade reservoirs. Great Basin Spadefoots tend to avoid sites occupied by breeding populations of either Western Toads or Western Tiger Salamanders, though in Deep Springs Valley, California, Great Basin Spadefoots and Black Toads do use the same breeding sites.

Great Basin Spadefoots continue to be common in suitable habitats; however, their patterns of abundance have been influenced by human activities. Throughout most of their range, springs and streams have been frequently dammed or diverted into ditches and impoundments, creating reservoirs where natural open water sources did not exist. In some cases, Great Basin Spadefoots have been able to capitalize on these changes and have become more abundant than before. But where urbanization, agriculture, and other land conversions have destroyed or harmed habitats, Great Basin Spadefoots have been extirpated or have declined.

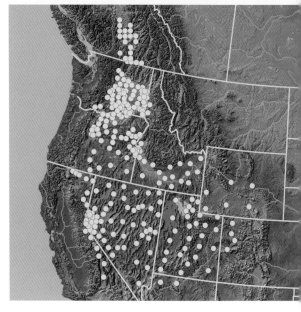

Great Basin Spadefoot

In Colorado, Great Basin Spadefoots are listed as a Species of Special Concern. In Canada, Great Basin Spadefoots are listed in British Columbia on the Blue List of Threatened and Endangered Species and as Threatened under the federal Species at Risk Act.

ORIGINAL ACCOUNT by Steven R. Morey; photograph by Gary Nafis.

Spea multiplicata (Cope, 1863)

Mexican Spadefoot

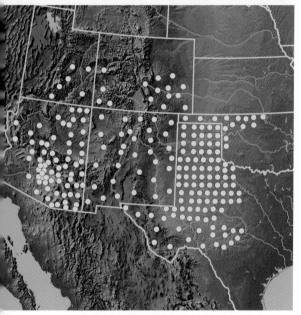

Mexican Spadefoot

Mexican Spadefoots are found from western Oklahoma and central Texas through New Mexico, southern Colorado, southeastern Utah, and Arizona and into Mexico (not shown).

Mexican Spadefoots occur in a wide range of arid and semiarid habitat types, such as grasslands, sagebrush flats, semiarid shrublands, river valleys, and agricultural lands. They are often found where the soil is sandy or gravelly.

Mexican Spadefoots confine their surface activity to short periods following summer showers and spend their time in shallow, soil-filled summer burrows, 1.3 to 10 cm deep. In July, or sometimes August, adults move from their underground refuges to reach breeding sites when rainfall creates temporary pools. Breeding sites include ephemeral pools and playas, tanks in rocky streambeds, isolated pools in temporary streams and arroyos, stock tanks, and pools that form at the base of road and railroad grades. During the rest of the year, Mexican Spadefoots dig down 20 to 90 cm into soil-filled winter burrows.

Urbanization, water projects, and irrigated agriculture have altered some natural habitats so severely that Mexican Spadefoots are less abundant than they were in the past; in some places, they have been eliminated. However, since they will use artificially created pools for breeding, they have been able to colonize some areas where suitable natural pools are rare or nonexistent.

In Colorado, Mexican Spadefoots are listed as a Species of Special Concern.

ORIGINAL ACCOUNT by Steven R. Morey; photograph by Brad Moon.

Xenopus laevis (Daudin, 1802)

African Clawed Frog

African Clawed Frogs are not native to North America, but they have been found living wild in many localities throughout the U.S., and apparently well-established, reproducing populations are known in Arizona, Texas, and southern California. These populations largely arose from animals that either had been released or had escaped from laboratory or pet stocks.

African Clawed Frog populations in the U.S. are found mainly in disturbed or artificial bodies of water, such as drainage ditches, flood control channels, golf course ponds, manmade lakes, irrigation canals, cattle tanks, and sewage plant effluent ponds. In their native habitats in sub-Saharan Africa, they are highly adaptable and will tolerate a wide variety of food availability, vegetation, substrate, turbidity, salinity, water temperature, and hydrology. Though African Clawed Frogs are almost wholly aquatic, they will travel overland when the ground is damp and so are capable of dispersing widely. The highest densities of African Clawed Frogs are reached in permanent, eutrophic, fish-free waters that have soft substrates and submerged vegetation and remain above 20°C for most of the year.

Efforts to eradicate African Clawed Frogs in California have not been successful. Concern over the potential impacts of introductions has led to banning the possession of African Clawed Frogs in several states, though they are still popular in the pet trade in many places.

African Clawed Frogs are considered to be an invasive species.

ORIGINAL ACCOUNT by John J. Crayon; photograph by Mike Redmer.

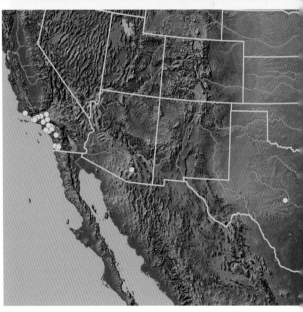

African Clawed Frog

Glandirana rugosa (Temminck and Schlegel, 1838)

Japanese Wrinkled Frog

Japanese Wrinkled Frog

Japanese Wrinkled Frogs were probably introduced to Hawaii from Japan over a century ago and are currently found on the islands of Kauai, Oahu, Maui, and Hawaii from elevations near sea level to at least 1100 m.

In Hawaii, Japanese Wrinkled Frogs breed in the pools and slow-moving waters of mountain streams and in lowland ponds. They occur in both pond and stream habitats. In smaller streams, they typically inhabit pools, whereas in larger streams with more current, they occur along the sides of quiet backwaters.

In Hawaii, Japanese Wrinkled Frogs are abundant and invasive in native forest. They have no legal protection.

ORIGINAL ACCOUNT by Fred Kraus; photograph by Pierre Fidenci.

Lithobates areolatus Baird and Girard, 1852

Crawfish Frog

Crawfish Frogs are found from southern and western Indiana, southern Illinois and south central Iowa south to eastern Texas and northwestern Louisiana in a range that generally encircles and excludes the Ozark Plateau and Ouachita Mountains except for the narrow band of the Arkansas River Valley through central Arkansas. Throughout their range, Crawfish Frogs have a discontinuous distribution, with populations localized in areas of suitable habitat.

In the northern parts of their range, Crawfish Frogs occupy prairies, reclaimed coal spoil grasslands, open wet woodlands, wooded valleys, river floodplains, pine forests, and meadows. Toward the south, though, Crawfish Frogs are largely limited to prairie, wet pastures, or grassland habitats, low-lying hay fields, and, occasionally, in woodland stream watersheds and river floodplains. Crawfish Frogs are also common in the hardpan, clay soil region of southern Illinois. Adult Crawfish Frogs are extremely fossorial, tied to abandoned crayfish or other small animal burrows they use as shelter. Burrows may exceed 1 to 1.5 m in depth, often have flattened platforms at the entrance, and may be located over 1 km from breeding wetlands.

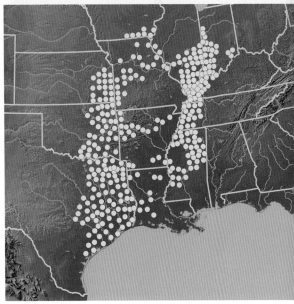

Crawfish Frog

Crawfish Frogs have been susceptible to regional extinction because they exhibit fidelity to burrows their. They are therefore susceptible to habitat fragmentation and cropland conversion. Agricultural practices may reduce or limit Crawfish Frog abundance by destroying sedentary adults and eliminating suitable burrowing habitats.

Crawfish Frogs are of considerable conservation concern and are listed as Endangered or In Decline in several states, including Iowa, Indiana, and Kansas.

ORIGINAL ACCOUNT by Matthew J. Parris and Michael Redmer; photograph by Mike Redmer.

Lithobates berlandieri Baird, 1854

Rio Grande Leopard Frog

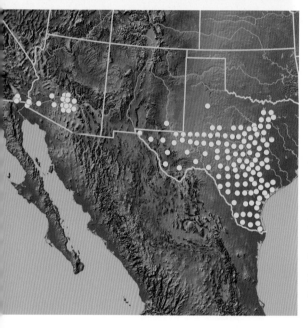

Rio Grande Leopard Frog

Rio Grande Leopard Frogs occur from central and western Texas and southeastern New Mexico, south along the Atlantic slope through at least southeastern Mexico (not shown). Rio Grande Leopard Frogs are also well established as an introduced species in Arizona on the Gila River drainage, along the Colorado River, near Yuma, Arizona, and in the Imperial Valley of southeastern California.

In their native range in Texas and New Mexico, Rio Grande Leopard Frogs are found in arid regions along streams or rivers and near cattle tanks, ponds, or ditches or in clear, flowing streams or permanent pools in intermittent stream drainages that originate from springs. In Arizona and California, Rio Grande Leopard Frogs are typically found on the edges of large slow-moving rivers; in agricultural ditches, drains, canals, and sumps; or even in a fish hatchery. The presence of holes or burrows where they can take refuge is important for Rio Grande Leopard Frogs. The numbers of Rio Grande Leopard Frogs at some sites can vary greatly from year to year and from season to season.

Rio Grande Leopard Frogs have no status under the U.S. Endangered Species Act, or in the states of Texas, New Mexico, Arizona, or California.

ORIGINAL ACCOUNT by James C. Rorabaugh; photograph by Mike Redmer.

Lithobates blairi Mecham, Littlejohn, Oldham, Brown and Brown, 1973

Plains Leopard Frog

Plains Leopard Frogs occur throughout much of the Great Plains and into the central Midwestern states of the U.S. Their range extends eastward from central New Mexico, central Colorado, and western Nebraska to central Indiana. From the northern limit of their range in southern South Dakota and central Iowa, Plains Leopard Frogs can be found south into northern Texas, excluding the Ozark Plateau of southern Missouri, Arkansas, and southeastern Oklahoma. Isolated populations occur in southern Illinois, New Mexico near Sierra Blanca, and southeastern Arizona on the western side of the Chiricahua Mountains and adjoining Sulfur Springs Valley. They occur at elevations from 110 to 2590 m.

Plains Leopard Frogs are to be found in prairie grasslands, oak savanna, and oak–pine forests. They are relatively drought- and heat-resistant, and adults are often associated with prairie potholes, pools in rocky canyons, live-stock tanks, streams, and irrigation ditches. During the summer, Plains Leopard Frogs can be found some distance from water, although they are seldom found in cultivated fields or in mature upland forests.

Plains Leopard Frogs are generally widespread and abundant throughout most of their range, although declines and extirpations of populations have been documented. In areas in Colorado now occupied by American Bullfrogs, Plains Leopard Frogs have become scarce.

In Arizona, Plains Leopard Frogs, are legally Protected, which means a permit is required to collect and/or possess them.

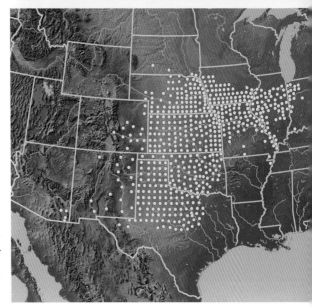

Plains Leopard Frog

ORIGINAL ACCOUNT by John A. Crawford, Lauren E. Brown, and Charles W. Painter; photograph by Mike Redmer.

Lithobates capito Le Conte, 1855

Gopher Frog

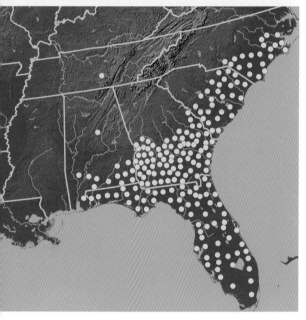

Gopher Frog

Historically, Gopher Frogs were distributed throughout the Gulf and Atlantic Coastal Plain from southeastern Alabama to North Carolina, with one isolated population known from the Ridge and Valley Province of Alabama and another in the Cumberland Plateau of Tennessee.

Gopher Frogs are associated with dry, fire-enhanced habitats, especially longleaf pine and turkey oak sandhills, pine flatwoods, sand pine scrub, and dry hammocks. Gopher Frogs are known to use the burrows of gopher tortoises for fire avoidance, predator protection, feeding, and/or escape from excessively cold or warm temperatures. Adults will also seek refuge in the burrows created by crayfish, oldfield mice, or pocket gophers, as well as within stump holes and the hollow interiors of previously submerged and partially decomposed willow tree branches.

There is concern about the status of Gopher Frogs in all the states in which they occur, but they are afforded legal protection only in North Carolina and Florida, where they are listed as of Special Concern, and in Alabama, where they are Protected. The U.S. Fish and Wildlife Service is currently evaluating their rangewide status to determine whether they warrant U.S. federal listing as an Endangered species.

ORIGINAL ACCOUNT by John B. Jensen and Stephen C. Richter; photograph by Dante Fenolio.

Lithobates catesbeianus Shaw, 1802

American Bullfrog

The native range of American Bullfrogs is from Nova Scotia south to central Florida and west across the Great Plains, probably including parts of northern Mexico (not shown). Although this historical range was large, American Bullfrogs were naturally limited in distribution by habitat, but accurate determination of the original distribution will be forever confused as a result of introductions outside the natural range in western U.S. and Canada. Many of these introductions began in the 1920s or 1930s.

Adult American Bullfrogs are mostly aquatic and prefer warm, lentic habitats such as vegetated shoals, sluggish backwaters and oxbows, farm ponds, reservoirs, marshes, and still waters with dead woody debris and dense and often emergent vegetation. They also occupy shorelines of lakes and streams. American Bullfrogs are often the predominant frog species in such permanent aquatic habitats, naturally occurring in proximity with species such as Green Frogs. Bullfrog tadpoles may transform into juvenile frogs in as little as a few months in the south of their range but, in northern populations, they may remain larvae for 3 or 4 years.

Large, voracious, and aggressive, introduced American Bullfrogs have been implicated in the decline or displacement of many native western North American species of frogs. American Bullfrogs may also be responsible for declines in Mexican Garter Snakes from the San Bernardino National Wildlife Refuge in California.

In contrast to their invasive spread and increasing abundance in the West, where they are introduced, American Bullfrogs appear to be declining in some populations in the East,

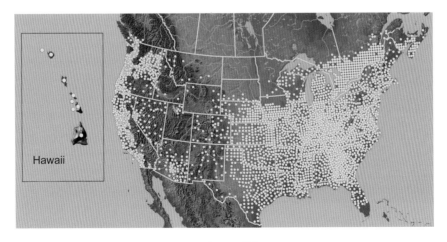

American Bullfrog

where they are native. American Bullfrogs were once common throughout the eastern U.S. and Canada but today, habitat loss and degradation, water pollution, pesticide contamination, and overharvesting are commonly invoked as causal factors in their decline. Local extirpations have been reported in southwestern Ontario.

American Bullfrogs are not listed as an endangered, threatened, or protected species under any provincial, state, or federal regulations. However, throughout its North American range, American Bullfrogs may be considered as regulated wildlife that can be harvested with either a hunting or a fishing license. Eastern states and provinces generally impose bag limits on harvesting of American Bullfrogs, but in western states and provinces, including California, Texas, Arizona, Idaho, Nevada, and British Columbia, where they are an introduced species, there may be no limits on harvesting American Bullfrogs.

ORIGINAL ACCOUNT by Gary S. Casper and Russ Hendricks; photograph by Scott Gillingwater.

Lithobates chiricahuensis Platz and Mecham, 1979[9]

Chiricahua Leopard Frog

Chiricahua Leopard Frog

Chiricahua Leopard Frogs are found in two separate areas of the southwest U.S. and adjacent Mexico. Northern montane populations are located along the Mogollon Rim, which forms the southern edge of the Colorado Plateau in central and eastern Arizona, and west central New Mexico. Southern populations, though, are located in the mountains and valleys south of the Gila River in southeastern Arizona and southwestern New Mexico and into Mexico along the eastern slopes of the Sierra Madre Occidental (not shown).

Chiricahua Leopard Frogs occupy a variety of natural and manmade, permanent to near-permanent, ponds, lakes, pools, and streams. Shallow-water localities with emergent and perimeter vegetation provide tadpoles and adults with basking sites, while deeper water, root masses, and undercut banks provide refuge from predators and likely serve as potential hibernacula during winter. The scattered occurrence of such habitats amid the overall arid landscapes of this region results in a fragmented range for Chiricahua Leopard Frogs.

Chiricahua Leopard Frogs appear to have disappeared from over 80% of historically known sites in both portions of its range.

Chiricahua Leopard Frogs are federally listed in the U.S. as Threatened.

ORIGINAL ACCOUNT by Michael J. Sredl and Randy D. Jennings; photograph by Brad Moon.

Lithobates clamitans Latreille, 1801

Green Frog

Green Frogs are found throughout most of the eastern U.S. and southeastern Canada from southern Ontario, Québec, and the Maritime Provinces south to the Gulf of Mexico. They occur almost everywhere east of central Minnesota south to central Oklahoma and eastern Texas. Green Frogs are absent from the southern half of Florida and from much of central Illinois. Isolated populations that were apparently introduced occur in western Iowa, northern Utah, and in several locations in the state of Washington and southwest British Columbia.

Green Frogs are found in most permanent aquatic habitats, including shorelines of lakes, marshes, swamps, streams, springs, and quarry and farm ponds, where emergent vegetation such as sedges, cattails, and rushes predominate. In northern populations, Green Frogs may spend 2 to 3 years as tadpoles before they transform. Adults usually go more than 1 m from water except on rainy nights. Green Frogs have been found associated with the mouths of caves and often will follow streams up and away from wetlands for great distances.

Green Frogs are relatively common throughout most of their range. Populations have undoubtedly been lost, as shorelines have been developed for recreational, business, and domestic uses.

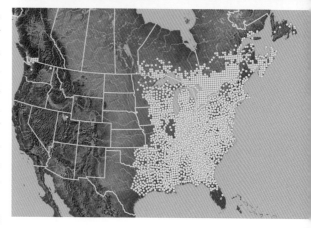

Green Frog

Green Frogs are classified as a Game Species in Missouri, Mississippi, Massachusetts, New York, New Jersey, Pennsylvania, and some other states, thereby providing them with protection in terms of hunting season and bag limits. In Kansas, Green Frogs are listed as a Threatened species. There are no regulations pertaining to Green Frogs in Canada.

ORIGINAL ACCOUNT by Thomas K. Pauley and Michael J. Lannoo; photograph by Mike Redmer.

Lithobates fisheri Stejneger, 1893[9]

Vegas Valley Leopard Frog

Once thought to be extinct, Vegas Valley Leopard Frogs were known only from the Las Vegas Valley of Nevada, where they were restricted to springs, seeps, and streams at elevations of about 610 m. A recent molecular analysis suggests that Vegas Valley Leopard Frogs still exist, represented by populations along the Mogollon Rim in central Arizona extending into western New Mexico previously thought to have been Chiricahua Leopard Frogs.

The isolated wetlands inhabited by Vegas Valley Leopard Frogs once were bordered by cottonwoods, willows, bulrushes, sedges, and cattails. The main drainage through the Las Vegas Valley, Las Vegas Creek, exhibited riparian vegetation only in an area about 13 km north of present-day Henderson and, typically, did not flow into the Colorado River. This suggests that aquatic habitats used by Vegas Valley Leopard Frogs were isolated from habitats used by other Leopard Frog species further east.

Because the habitat of Vegas Valley Leopard Frogs was restricted, the status of these animals has always been considered to be precarious. As the population of Las Vegas grew, downstream flows in Las Vegas Creek were augmented by secondary sewage effluent and irrigation runoff, whereas the upstream, headwater springs were used progressively to supply water for the various needs of a burgeoning town.

Vegas Valley Leopard Frog

Until recently, Vegas Valley Leopard Frogs were considered to be extinct and, so far, have no other official conservation status under that name. However, the Mogollan Rim populations are federally listed in the U.S. as Threatened, as "Chiricahua Leopard Frogs."

ORIGINAL ACCOUNT by Randy D. Jennings; photograph by Randy Jennings.

Lithobates grylio Stejneger, 1901

Pig Frog

Pig Frogs are endemic to the southeastern Coastal Plain of the U.S. from southern South Carolina to the Everglades of Florida and west along the Gulf Coast to extreme southeastern Texas. Pig Frogs have also been found on barrier islands off the Atlantic Coast, though that is outside of their native range. They have been introduced on Andros and New Providence islands in the Bahamas and are well established in northern Puerto Rico (not shown).

Pig Frogs are largely aquatic, typically remaining within permanent wetland habitats throughout the year. Pig Frogs will occupy the lodges of round-tailed muskrats in Florida.

Pig Frogs have been and still are considered common throughout their range. In Florida, Pig Frogs are the second most abundant frog, and the one most commonly gigged by hunters. Populations of Pig Frogs do not appear to have diminished. On the contrary, their geographic distribution has expanded, and Pig Frogs, unlike most other anuran species, appear to be positively affected by residential development.

Pig Frogs are not considered Threatened, Endangered, or of Special Concern by any state or federal agency. In Louisiana, a state fishing license is required to take Pig Frogs during a limited hunting season.

ORIGINAL ACCOUNT by Stephen C. Richter; photograph by Brad Moon.

Pig Frog

Lithobates heckscheri Wright, 1924

River Frog

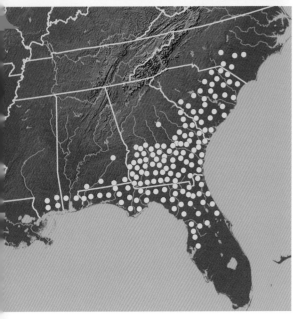

River Frog

River Frogs are found along the Atlantic and Gulf Coastal Plains of the U.S. from south central North Carolina south to the Oklawaha River, Florida, and west to southern Mississippi.

River Frogs inhabit the swampy edges of rivers and streams and can be found at night in shrubbery and the bases of trees along banks. They also occur along the edges of shallow impoundments, such as beaver ponds, associated with vegetation such as titi, bay, cypress, or sphagnum moss. River Frogs are nocturnal, terrestrial, and associated with emergent vegetation.

River Frogs are locally distributed and are perhaps the least known ranid frog of the southeastern U.S. A notable disparity in relative abundance between adults and juveniles in this species indicates an extremely low rate of survival to adulthood.

In North Carolina, River Frogs were last documented in 1975, and they are considered to be a species of Special Concern in that state.

ORIGINAL ACCOUNT by Brian P. Butterfield and Michael J. Lannoo; photograph by Aubrey Huepel.

Lithobates okaloosae Moler, 1985

Florida Bog Frog

Florida Bog Frogs are known only from northwestern Florida, where they occur along small streams draining to Titi Creek, the East Bay River, or the lower Yellow River, all of which ultimately drain to Escambia Bay. The Titi Creek populations appear to be isolated by approximately 30 km from the nearest other populations in the lower Yellow River basin. Most known populations of Florida Bog Frogs occur on Eglin Air Force Base.

Florida Bog Frogs are associated with shallow, nonstagnant, slightly acidic seeps and shallow, boggy overflows of seepage-fed streams, usually in association with lush beds of sphagnum moss.

Florida Bog Frogs were unknown prior to 1982. Nothing is known about their historical distribution, but there is no reason to believe that it was substantially greater than it is at present.

Florida Bog Frogs are considered to be a Species of Special Concern in Florida.

ORIGINAL ACCOUNT by Paul E. Moler; photograph by Patrick Gault.

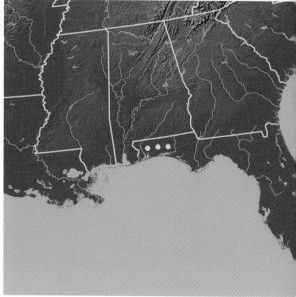

Florida Bog Frog

Lithobates onca Cope, 1875

Relict Leopard Frog

Relict Leopard Frog

Relict Leopard Frogs have been known occur at elevations below 600 m in the Virgin River drainage from the vicinity of Hurricane in Utah to the Overton Arm of Lake Mead in Nevada, along the Muddy River in Nevada, and from the Black Canyon along the Colorado River downstream from Lake Mead. Currently, they remain only in the Overton Arm and the Black Canyon.

Relict Leopard Frogs inhabit permanent warm springs, streams, and spring-fed wetlands, where water temperatures remain between 30 and 55°C. Adults prefer relatively open shorelines without dense vegetation and are usually found at the water›s edge or, occasionally, in adjacent low riparian vegetation.

Relict Leopard Frogs are rare. Populations of Relict Leopard Frogs have been extirpated throughout the species' range in tandem with the elimination or dramatic alteration of aquatic habitats due to agriculture, marsh draining, and water development. The introduction and spread of American Bullfrogs, crayfish, and predaceous fishes may also have been partly responsible for declines of Relict Leopard Frog populations.

Relict Leopard Frogs are currently designated by the U.S. Fish and Wildlife Service as a candidate for listing under the Endangered Species Act.

ORIGINAL ACCOUNT by David F. Bradford, Randy D. Jennings, and Jef R. Jaeger; photograph by Gary Nafis.

Lithobates palustris LeConte, 1825

Pickerel Frog

Pickerel Frogs are distributed from Nova Scotia and New Brunswick; west through southern Québec and Ontario, to Michigan, Wisconsin, and extreme southeastern Minnesota; southwest to the Gulf Coast in eastern Texas; and throughout most of the eastern U.S. They are absent, though, from much of the eastern Gulf Coastal Plain, the Florida Peninsula, and the interior prairies of Illinois.

Pickerel Frogs prefer streams or ponds with cool, unpolluted water with dense herbaceous vegetation, though some coastal and floodplain populations are reported to occupy swamps. Pickerel Frogs are the most cave-adapted North American anuran, and they are often abundant in areas of karst topography. They may also enter abandoned mines. Other habitats used by Pickerel Frogs include wooded wetlands, bogs, and shrubby, open meadows. Pickerel Frogs may be less abundant where stream bank vegetation is mowed or grazed and may be absent from areas that have been logged.

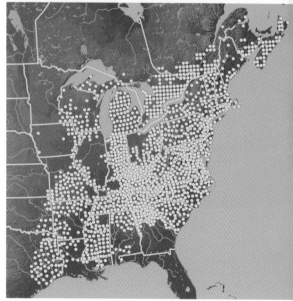

The somewhat specialized habitat characteristics of Pickerel Frogs, and their intolerance of pollution, may make them vulnerable to human activities. They are common in some regions but variously rare, uncommon, or localized in other parts of their range.

Pickerel Frogs are listed as Declining in Iowa, as Declining and a Species of Special Concern in Wisconsin, and a Species of Special Concern in

Pickerel Frog

Minnesota. Pickerel Frogs may be extirpated in Kansas. In Québec, they are considered likely to be designated endangered (*menacée*) or threatened (*vulnérable*).

ORIGINAL ACCOUNT by Michael Redmer; photograph by Scott Gillingwater.

Lithobates pipiens Schreber, 1782

Northern Leopard Frog

Northern Leopard Frogs occur from Newfoundland and southern Québec, south through New England to West Virginia, and west across the Canadian provinces and northern and central portions of the U.S. to British Columbia, Oregon, Washington, and northern California. Populations range south to southern Nevada, central Arizona, and in southern New Mexico and extreme southwestern Texas to near the Mexican border. The species has been introduced at Lake Tahoe, California, western Newfoundland, Vancouver Island, and elsewhere. In the southwestern states, the species generally occurs at higher elevations, such as the mountains of northern and central Arizona and New Mexico.

Adult Northern Leopard Frogs require shallow, well-vegetated wetland habitats such as marshes for breeding, deep; well-oxygenated ponds or streams for overwintering; and relatively open upland areas as postbreeding habitats for summer foraging. These uplands are typically grassy areas, meadows, or fields but can include sites such as peat bogs and fields sown with perennial forage crops such as grass, alfalfa, and clover. Postbreeding summer habitats usually do not include barren ground, open sandy areas, heavily wooded areas, heavily cultivated fields (especially

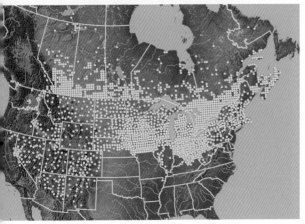

Northern Leopard Frog

those that have been cut recently), heavily grazed pastures, or closely mowed lawns. Northern Leopard Frogs in the relatively arid landscapes of southwestern Alberta may be limited by lack of suitable upland postbreeding habitats and lack of dispersal opportunities, since it is vital that all three types of habitat (breeding, overwintering, and summer foraging) are connected and accessible by the frogs.

Although Northern Leopard Frogs currently occur throughout most of their historical range, population declines and loss since the 1960s or earlier have resulted in their extirpation from some areas, particularly in the western portions of their range.

Northern Leopard Frogs are federally listed as Endangered in British Columbia and a species of Special Concern in the Canadian Prairie Provinces under the Canadian Species at Risk Act, but populations in eastern Canada are not considered to be at risk. Northern

Leopard Frogs have no status under the U.S. Endangered Species Act, although western populations have been recently proposed for listing. States and provinces often have species designations that include Northern Leopard Frogs or protect the species from hunting. For example, the species is provincially listed as Threatened in Alberta and in on the British Columbia Red List of Threatened and Endangered Species.

ORIGINAL ACCOUNT by James C. Rorabaugh; photograph by Scott Gillingwater.

Lithobates septentrionalis (Baird, 1854)

Mink Frog

Mink Frogs occur from southern Labrador and the Maritime Provinces across central Québec to James Bay, through northern Ontario to Minnesota and southeastern Manitoba, and south to northern New York and northern Wisconsin, with isolated populations in northern Québec and northern Labrador. The southern limit of distribution of Mink Frogs, at 43°N latitude, is the most northerly of any North American anuran.

Mink Frogs are highly aquatic and typically occupy rivers, lakes, ponds, pools, puddles, ditches, and streams. They avoid waters with rapid currents and large wave activity, preferring quiet bays and protected areas with a high abundance of aquatic vegetation, especially floating water lilies and pickerel weed, or mats of sphagnum moss. Mink Frogs are substantially less tolerant of desiccation and more tolerant of hydration than other northern anurans. Terrestrial activity is usually restricted to periods of nocturnal precipitation. Mink frog tadpoles overwinter at least once before they transform.

Mink Frogs generally are considered to be locally common in suitable habitats. Mink Frogs benefit from the activities of beavers in creating ponds, and recent increases in beaver abundance may have increased the availability of their breeding habitats in some areas.

Mink Frog

Mink Frogs are not listed under any state, provincial, or federal regulations.

ORIGINAL ACCOUNT by Gary S. Casper; photograph by Tony Gamble.

Lithobates sevosus Goin and Netting, 1940

Dusky Gopher Frog

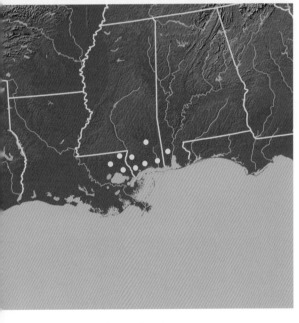

Dusky Gopher Frog

Dusky Gopher Frogs are endemic to the Gulf Coastal Plain in Louisiana, Mississippi, and southwestern Alabama, but though they were once abundant, Dusky Gopher Frogs are known currently from only a few breeding populations in the DeSoto National Forest region of Mississippi. They are thought to now be extirpated from both Louisiana and Alabama.

Outside of the breeding season, a typical habitat for Dusky Gopher Frogs consists of both upland and flatland longleaf pine forests with relatively open canopies. The adult frogs reside in underground retreats associated with gopher tortoise or small mammal burrows, stump holes, and root mounds of fallen trees. Dusky Gopher Frogs often can be seen outside of these retreats during the day. Dusky Gopher Frogs are threatened primarily by human-induced stressors.

Dusky Gopher Frogs are listed as Endangered under the U.S. Endangered Species Act.

ORIGINAL ACCOUNT by Stephen C. Richter and John B. Jensen; photograph by Mike Redmer.

Lithobates sphenocephalus Cope, 1886[10]

Southern Leopard Frog

Southern Leopard Frogs are distributed throughout the southeastern quarter of the continental U.S., exclusive of the Appalachian Highlands. Their range extends from central Texas and Oklahoma and eastern Kansas eastward to the Atlantic Coast and north to southeastern New York, including Long Island.

Southern Leopard Frogs use all types of shallow freshwater habitats, including temporary pools, cypress ponds, ponds, lakes, ditches, irrigation canals, and stream and river edges. They will inhabit slightly brackish coastal wetlands, too, as they are able to tolerate salinities up to 2.14 parts per thousand. Southern Leopard Frogs will move into terrestrial habitats to feed during the summer, when vegetation in pastures, fields, and sod lands affords shade and shelter. They are primarily nocturnal. Annual periods of activity range from nearly year-round in the south to winter dormancy in the north.

Southern Leopard Frogs are considered the most abundant and ubiquitous frogs in Florida and Alabama. In Tennessee, Southern Leopard Frogs are common throughout much of this state, but are apparently limited to the east by the higher elevations of the Blue Ridge Mountains.

Southern Leopard Frogs are not considered a species of conservation concern by any state or by the U.S. government.

ORIGINAL ACCOUNT by Brian P. Butterfield, Michael J. Lannoo, and Priya Nanjappa; photograph by Mike Redmer.

Southern Leopard Frog

Lithobates sylvaticus LeConte, 1825

Wood Frog

Wood Frogs are the most widespread North American amphibian species. They occur from the southern Appalachian Mountains of Georgia, north through eastern Canada to Ungava Bay, and west across the boreal forest region to the Bering Sea in Alaska. The western edge of their range runs roughly diagonally from Alaska southeast through British Columbia and across the Canadian prairies most of North Dakota to run through the Upper Midwestern states of Minnesota, Wisconsin, Illinois, and Indiana, to northeastern Alabama. Disjunct populations occur in Colorado, Wyoming, Idaho, and the Ozark Plateau.

Wood Frogs are found in a variety of habitats, including tundra, subalpine woodlands, willow thickets, wet meadows, bogs, and both coniferous and deciduous temperate forests dominated by various different tree species. They breed in a wide range of permanent and ephemeral bodies of water. Wood Frogs survive subfreezing conditions as they hibernate at or near the soil surface during winter by flooding their blood systems with glucose as they freeze. This protects their bodies from frost damage until they thaw. Throughout much of

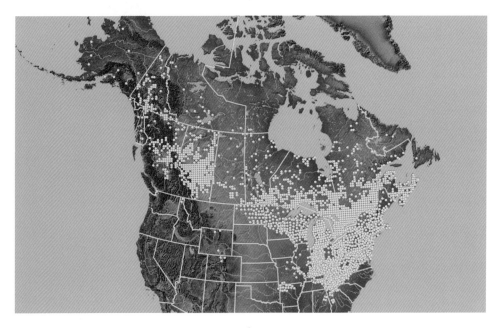

Wood Frog

their range, Wood Frogs are a common and familiar species; in the far north of their range, they are the only amphibian present.

Wood Frogs are sensitive to the edge effects and reduced canopy cover created by forest harvesting but can persist in semiurban and agricultural settings in parts of their range.

In several states along the southern periphery of their range, including Alabama, Arkansas, Colorado, Illinois, Missouri, and New Jersey, Wood Frogs are either considered to be Restricted, Uncommon, or Rare, or they are afforded legal protection.

ORIGINAL ACCOUNT by Michael Redmer and Stanley E. Trauth; photograph by Jim Harding.

Lithobates tarahumarae Boulenger, 1917

Tarahumara Frog

Tarahumara Frogs are known from montane canyons in extreme southern Arizona south into northern Mexico (not shown).

Tarahumara Frogs inhabit oak and pine–oak woodland, where there are moist refugia among rocks and boulders along streams and at plunge pools that they can use as winter retreats. Permanent water is necessary for larval metamorphosis. Tarahumara Frogs may also inhabit artificial impoundments.

Tarahumara Frogs were known historically from six localities north and west of Nogales, Arizona, but they have been extirpated from all localities in Arizona, having last been seen in the U.S. in 1983.

Tarahumara Frogs are considered to be an Endangered species by the Arizona Game and Fish Department and are included on a draft state list of Species of Concern. However, Tarahumara Frogs have no status under the U.S. Endangered Species Act.

ORIGINAL ACCOUNT by James C. Rorabaugh and Stephen F. Hale; photograph by Bruce Taubert.

Tarahumara Frog

Lithobates virgatipes Cope, 1891

Carpenter Frog

The range of Carpenter Frogs extends along the Atlantic Coastal Plain from southern New Jersey to southeastern Georgia and into northeastern Florida to about the Santa Fe River.

Carpenter Frogs breed in permanent, highly vegetated, low- to high-acidity sphagnum ponds, beaver ponds, freshwater marshes, interdunal cypress swales, and pocosins. Adults do not venture far from water and apparently remain in wetlands in all seasons. Carpenter Frogs may be common locally in some areas, but most populations consist of small numbers of individuals.

Carpenter Frogs are listed as a Species of Special Concern in Virginia because of the low number of locations in which they are found within the state.

ORIGINAL ACCOUNT by Joseph C. Mitchell; photograph by Mike Redmer.

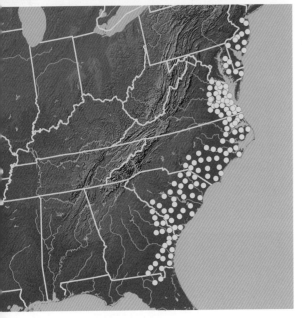

Carpenter Frog

Lithobates yavapaiensis Platz and Frost, 1984

Lowland Leopard Frog

Lowland Leopard Frogs were distributed from northwestern Arizona through central and southeastern Arizona, southwestern New Mexico, and into adjacent northern Mexico (not shown). Populations also were known along the lower Colorado River and in the Coachella Valley in southwestern Arizona and southeastern California.

Lowland Leopard Frogs inhabit aquatic systems in desert scrub to pinyon–juniper woodland. They are habitat generalists and breed in a variety of natural and manmade aquatic systems, including rivers, permanent streams, permanent pools in intermittent streams, beaver ponds, springs, earthen cattle tanks, canals, irrigation sloughs, abandoned swimming pools, and ornamental backyard ponds. Lowland Leopard Frogs tend to be concentrated at springs, near debris piles, at heads of pools, and near deep pools associated with root masses. Thermal springs, where constant flow and warm water temperatures permit year-round adult activity and winter breeding, are particularly important breeding habitat for Lowland Leopard Frogs in New Mexico.

Lowland Leopard Frog populations appear vulnerable to large-scale mortality, on a frequent basis, due to drought, flooding, disease and/or deteriorated water conditions.

Lowland Leopard Frogs were formerly considered to be a Category 2 candidate species for listing under the U.S. Endangered Species Act but currently are not listed under any state or federal regulations in the U.S.

ORIGINAL ACCOUNT by Michael J. Sredl; photograph by Ken Dodd.

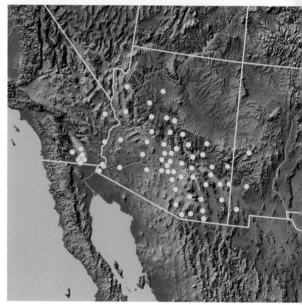

Lowland Leopard Frog

Rana aurora Baird and Girard, 1852

Northern Red-legged Frog

Northern Red-legged Frogs occur along the Pacific Coast from northern California to the midcoast of British Columbia between elevations of sea level and 1200 m, including Vancouver Island and islands in Georgia Strait. There are introduced populations on the Queen Charlotte Islands of British Columbia and on Chichagof Island in Alaska.

Adult Northern Red-legged Frogs generally leave breeding sites relatively soon after the breeding period and may move substantial distances upstream from breeding pools through wet forests and riparian areas. The summer habitats of adults include streambanks and moist riparian areas adjacent to standing water.

Throughout much of their lowland range, Northern Red-legged Frogs may be displaced by nonnative American Bullfrogs, which are either outcompeting the native Northern Red-legged Frogs or moving into disturbed and altered habitats that are no longer preferred by the native species. In some places, Northern Red-legged Frog tadpoles have been found to compete poorly against American Bullfrog tadpoles when food resources are concentrated. Northern Red-legged Frogs also respond negatively to the presence of introduced, nonnative fish in their breeding habitats. In Oregon's Willamette Valley, the reduced and fragmented present distribution of Northern Red-legged Frogs is likely the result of intensive land use and the establishment of a variety of such nonnative predators.

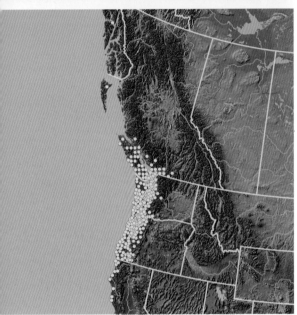

Northern Red-legged Frog

Northern Red-legged Frogs are considered a Species of Special Concern in California, Sensitive-Vulnerable in Oregon's Willamette Valley, and Sensitive-Unknown elsewhere in Oregon. They are listed as a species of Special Concern under Canada's Species at Risk Act and on British Columbia's Blue List of Threatened and Endangered Species.

ORIGINAL ACCOUNT by Christopher A. Pearl; photograph by William Leonard.

Rana boylii Baird, 1854

Foothill Yellow-legged Frog

Foothill Yellow-legged Frogs are native to much of southwestern Oregon west of the crest of the Cascade Mountains and south through California in the mountains and foothill regions encircling the Central Valley as far south as the San Gabriel River system. Foothill Yellow-legged Frogs range from elevations near sea level to 1940 m in California.

Foothill Yellow-legged Frogs are primarily stream dwellers and are found mostly near flowing water with rocky substrate riffles containing cobble-sized or larger rocks, which can be used as egg-laying sites. The adult Foothill Yellow-legged Frogs sit on open, sunny banks.

Foothill Yellow-legged Frogs are susceptible to a wide range of environmental impacts, including loss of habitat, pesticides, competition, and predation from nonnative species such as warm-water predatory fish, American Bullfrogs, crayfish, disease, water impoundments, logging, mining, and grazing in riparian zones. Dams along many river drainages negatively impact Foothill Yellow-legged Frogs. Although healthy populations of Foothill Yellow-legged Frogs are scattered throughout the central western portions of its range, populations in the southern Sierra Nevada foothills are unlikely to remain viable for much longer. In Oregon, there is only one known population in the Cascade Foothills on the east side of the Willamette Valley, and Foothill Yellow-legged Frogs are rare in the Klamath Basin.

California lists Foothill Yellow-legged Frogs as a species of Special Concern,

ORIGINAL ACCOUNT by Gary M. Fellers; photograph by William Leonard.

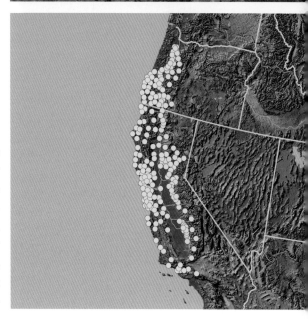

Foothill Yellow-legged Frog

Rana cascadae Slater, 1939

Cascades Frog

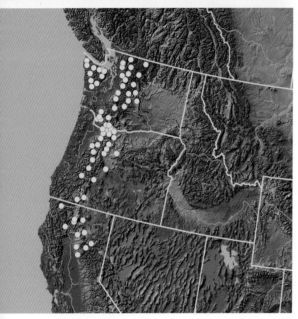

Cascades Frog

Cascades Frogs are found in middle- and high-elevation ponds and lakes at elevations ranging from 400 to 2500 m throughout the Cascades Range from the very northern edge of California's Sierra Nevada to within 25 km of the British Columbia border. Cascades Frogs also occur in the Olympic Mountains of Washington and in the Trinity Alps, Mt. Shasta, and Mt. Lassen areas of northern California, but they have been extirpated from the majority of their historical localities in California.

Cascades Frogs are generally associated with open wetland habitats and commonly occupy moist meadows and relatively small permanent and temporary ponds. They are also found along streams in summer, especially at lower elevations, where lentic habitats are less common. Cascades Frogs generally stay close to water, particularly along sunny shores, under dry summer conditions but will traverse uplands during high humidity.

Declines in populations of Cascades Frogs are well documented in the southern portion of their range but not so much to the north. Their abundance and distribution is negatively impacted by the presence of introduced salmonid fishes. In the northern and central Oregon Cascade Range, in Olympic and Mount Rainier national parks, and in the Mt. St. Helens Volcanic Monument in Washington, Cascades Frogs remain widespread and common.

Cascades Frogs are considered a Species of Special Concern in California, and Sensitive–Vulnerable in Oregon.

ORIGINAL ACCOUNT by Christopher A. Pearl and Michael J. Adams; photograph by William Leonard.

Rana draytonii Baird and Girard, 1852

California Red-legged Frog

California Red-legged Frogs are native to the Sierra Nevada foothills and Coast Range of California, circling the Central Valley, and south through northwestern Baja California in Mexico (not shown). Two isolated, introduced populations of California Red-legged Frogs occur near to each other in Nevada.

California Red-legged Frogs are primarily pond frogs, but they also inhabit marshes, streams, and lagoons during the breeding season. During other parts of the year, they will occupy areas that stay moist and cool within 2 to 3 km of a breeding site. Such sites include coyote bush, California blackberry thickets, and root masses associated with willow and California bay trees. Even a 1- or 2-meter wide coyote bush thicket along a tiny, intermittent creek surrounded by heavily grazed grassland will be enough to support California Red-legged Frogs.

Although California Red-legged Frogs were once abundant throughout much of California, they are nearly extirpated in both the Sierra Nevada foothills and in the southern quarter of its range. In the central California Coast Range, the number of extant populations of California Red-legged Frogs has been reduced substantially by the loss of suitable habitat. In California south of Santa Barbara, there are only two extant populations.

Much of the prime habitat for California Red-legged Frogs has been converted to other land uses, especially in the Los Angeles and San Francisco Bay areas and in the Sierra foothills. The most serious threats to California Red-legged Frogs are loss of habitat from urbanization and agriculture and exposure to pesticides.

California Red-legged Frogs are listed in the U.S. as a Threatened species under the Endangered Species Act.

ORIGINAL ACCOUNT by Gary M. Fellers; photograph by William Leonard.

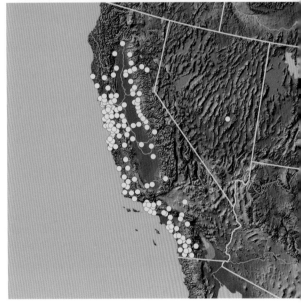

California Red-legged Frog

Rana luteiventris Thompson, 1913

Columbia Spotted Frog

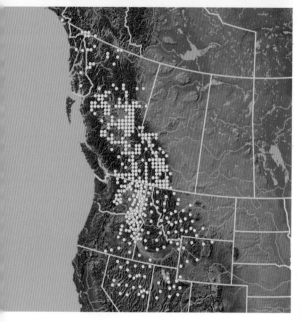

Columbia Spotted Frog

Columbia Spotted Frogs range from extreme southeastern Alaska south through British Columbia and southwestern Alberta, western Montana and Wyoming, northern and central Idaho, eastern Oregon and Washington at elevations up to 2600 m. Isolated populations of Columbia Spotted Frogs exist at middle to high elevations in parts of the Great Basin in Utah and Nevada.

Columbia Spotted Frogs frequent permanent or ephemeral pools or flowing wetlands ranging in size from ditches to lakes, as well as moist meadows. They prefer sites with floating and/or emergent vegetation and, in some populations, are found within dense willow clumps or among talus. Columbia Spotted Frogs will also use human-created wetlands. After breeding, adults may move away from breeding sites into surrounding habitats that range from mixed coniferous and subalpine forests to arid grass and brushlands. This reduces their risk of predation by garter snakes that gather to feed on tadpoles and newly metamorphosed animals.

Across much of their range from Idaho, western Montana, eastern Oregon, and northwestern Wyoming north through British Columbia, Columbia Spotted Frogs are common and usually abundant in many areas, especially in fishless lakes and ponds. However, in the southern parts of their range in eastern Oregon, northern Nevada, and western Utah, Columbia Spotted Frog populations are rare, small, and very isolated.

Columbia Spotted Frogs in the Great Basin of Nevada and southern Idaho and in the Wasatch Front and West Desert of Utah are a candidate species for protection under the U.S. Endangered Species Act.

ORIGINAL ACCOUNT by Jamie K. Reaser and David S. Pilliod; photograph by William Leonard.

Rana muscosa Camp, 1917

Southern Mountain Yellow-legged Frog

Southern Mountain Yellow-legged Frogs are native to the mountains of southern California, including at elevations between 1600 to 3790 m the southern Sierra Nevada and from 300 to 2300 m on Palomar Mountain and in the San Jacinto, San Bernardino, and San Gabriel Mountains ringing Los Angeles.

Southern Mountain Yellow-legged Frogs occupy diverse aquatic habitats, including streams in narrow rock-walled canyons, streams in the chaparral belt in southern California and granitic glacially carved lakes in the Sierra Nevada. They are most often seen on a wet substrate within 1 m of the water's edge.

Southern Mountain Yellow-legged Frog populations have declined nearly to extinction. Over 95% of known populations have been extirpated and those that remain are very small, usually having fewer than 100 adults. Southern Mountain Yellow-legged Frogs are extinct on Palomar and Breckenridge Mountains and were thought to be extinct from the San Bernardino Mountains until a recent discovery of a small population.

Southern Mountain Yellow-legged Frogs are listed as Endangered under the distinct population clause of the U.S. Endangered Species Act and are considered as a species of Special Concern in California.

ORIGINAL ACCOUNT by Vance Vredenburg, Gary M. Fellers, and Carlos Davidson; photograph by Gary Nafis.

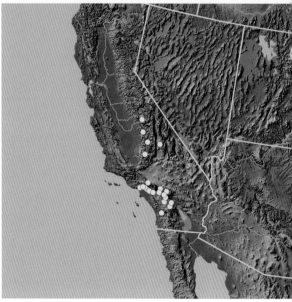

Southern Mountain Yellow-legged Frog

Rana pretiosa Baird and Girard, 1853

Oregon Spotted Frog

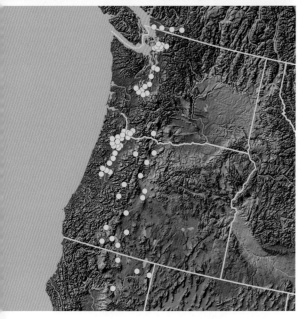

Oregon Spotted Frog

Oregon Spotted Frogs historically ranged west of the Cascade crest from northeastern California north to southwestern British Columbia. Oregon Spotted Frogs occur in largely isolated populations within larger emergent wetlands at elevations from sea level to 1635 m, with maximum elevations increasing in the southern portion of their range.

Adult Oregon Spotted Frogs are highly aquatic and are rarely found more than 2 m from surface water. Adults prefer microhabitats characterized by deeper water and more open forest canopy and remain associated with standing water throughout the summer season. They may also use seasonally temporary pools within 300 m of the larger, permanent wetlands they use as breeding sites.

Oregon Spotted Frogs have been extirpated from 70 to 90% of their native range, including their entire range in northeastern California. They are gone from across their historical lowland range in western Oregon and only five extant sites are known from western Washington and British Columbia.

In the U.S., Oregon Spotted Frogs are considered to be Endangered by the State of Washington, Sensitive–Critical in Oregon, and a Species of Special Concern in California, where they have been recommended for Endangered status. They are a candidate species for listing by the U.S. Fish and Wildlife Service. In Canada, Oregon Spotted Frogs are on the British Columbia Red List of Threatened and Endangered Species and are listed as Endangered under Canada's Species at Risk Act.

ORIGINAL ACCOUNT by Christopher A. Pearl and Marc P. Hayes; photograph by William Leonard.

Rana sierrae Camp, 1917

Sierra Nevada Yellow-legged Frog

Sierra Nevada Yellow-legged Frogs occur at high elevations in the Sierra Nevada Range of California and extreme western Nevada from the Diamond Mountains south to Matlock Lake.

Sierra Nevada Yellow-legged Frogs are highly aquatic and are always found within 1 to 2 m from the edge of water. Breeding begins soon after ice-off or early in spring. The length of the larval stage depends upon elevation. At lower elevations, where the summers are longer, tadpoles are able to grow to metamorphosis in a single season. At higher elevations, tadpoles must overwinter at least once and may take 2 to 4 years of growth before they are large enough to transform.

Sierra Nevada Yellow-legged Frogs have declined dramatically despite the fact that most of their habitat is protected in National Parks and National Forest lands. They are now extinct in the state of Nevada, and over 92% of known populations have been extirpated in California. The two most important factors leading to these declines are disease and introduced predators, particularly stocked salmonid fish.

Despite their steep decline, Sierra Nevada Yellow-legged Frogs are not listed under any state or federal regulations in the U.S.

ORIGINAL ACCOUNT by Vance Vredenburg; photograph by Gary Nafis.

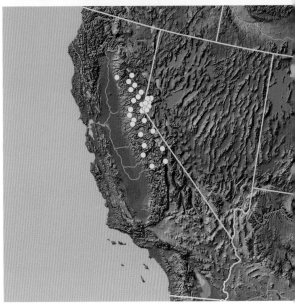

Sierra Nevada Yellow-legged Frog

Rhinophrynus dorsalis Duméril and Bibron, 1841

Burrowing Toad

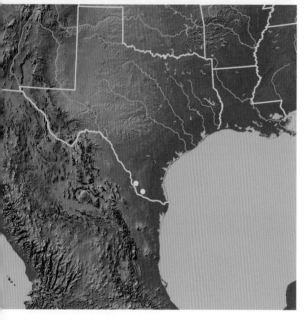

Burrowing Toad

In the U.S., Burrowing Toads have been found only in extreme southwestern Texas, but their range extends south from there through Mexico and Central America to Costa Rica (not shown). Burrowing Toads occur at elevations from sea level to 600 m.

Burrowing Toads have been a well known, if infrequently encountered, burrowing anuran of lowland coastal areas in Mesoamerica for some 150 years, but were not discovered as a component of the U.S. herpetofauna until 1964, when breeding populations were found in southern Texas. Burrowing Toads surface only to breed during heavy rains, perhaps only once per year, and thus they are encountered infrequently by humans.

Burrowing Toads are typically found in arid or savanna, tropical or subtropical, nonforested habitats with friable soils. They can also be found in cultivated fields and gardens. Burrowing Toads will breed any time of the year after rains create sufficiently deep breeding pools. Individuals occasionally are encountered on roads when there is breeding activity nearby.

Because these are secretive animals, determining their distribution and abundance, and therefore their conservation status, is a problem.

Burrowing Toads are considered a Threatened species in Texas but are not listed by the U.S. government.

ORIGINAL ACCOUNT by M.J. Fouquette Jr.; photograph by Dante Fenolio.

SALAMANDERS OF NORTH AMERICA

Ambystoma annulatum Cope, 1886

Ringed Salamander

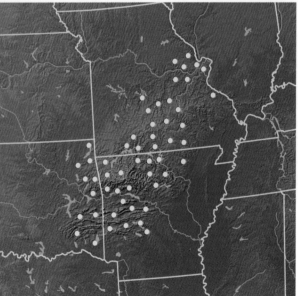

Ringed Salamander

Ringed Salamanders are endemic to the Ozark and Ouachita Mountains of Arkansas, Missouri, and Oklahoma.

Ringed Salamanders are found in forested areas under logs, leaves, and rocks or burrowed into the soil; they are seldom found in the open. They typically breed in fishless, woodland pools or shallow ponds, though they may use farm ponds, even those heavily used by livestock in open pastures. During September to early November, adults are stimulated to migrate at night to breeding ponds by medium to heavy rains and cool temperatures.

The distribution of Ringed Salamanders appears to be stable, and some populations retain large numbers of breeding adults. They may, however, be subject to illegal harvest as fish bait.

Ringed Salamanders are not listed in any of the states where they occur, nor are they federally listed in the U.S.

ORIGINAL ACCOUNT by Stanley E. Trauth; photograph by Dante Fenolio.

Ambystoma barbouri Kraus and Petranka, 1989

Streamside Salamander

Streamside Salamanders occur in central and western Kentucky, central Tennessee, southeastern Indiana, southwestern Ohio, and extreme western West Virginia along the Ohio River.

Streamside Salamanders are usually found in upland deciduous forests and are most common in regions with exposed limestone. They are extremely fossorial, and above-ground activity is observed mostly during breeding migrations on rainy nights in first- and second-order streams. Streamside Salamanders have an extended breeding season that lasts from late fall to early spring.

Streamside Salamanders usually are not found in streams where surrounding forested land has been cleared, suggesting that deforestation and development around streams and ravines within their range are detrimental to this species.

ORIGINAL ACCOUNT by Mark B. Watson and Thomas K. Pauley; photograph by Kevin Messenger.

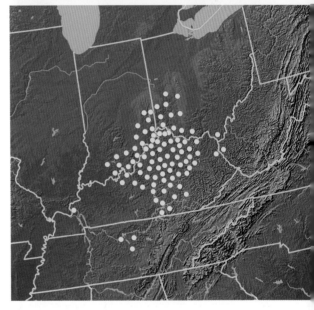

Streamside Salamander

Ambystoma bishopi Goin, 1950

Reticulated Flatwoods Salamander

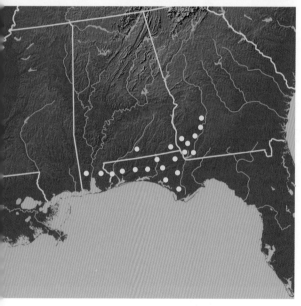

Reticulated Flatwoods Salamanders are restricted to the Coastal Plain of the southeastern U.S. from the Apalachicola and Flint rivers in the central portion of the Florida Panhandle and southwestern Georgia into southern Alabama.

Reticulated Flatwoods Salamanders breed in small, shallow, ephemeral ponds dominated by pond cypress and water tupelo. Females deposit their eggs singly in small groups in dry pond basins in the open on bare soil; under cover beneath logs, leaf litter, dead grass, or sphagnum mats; or within crayfish burrows. Larvae hatch when rains fill the ponds. Adult habitats include longleaf pine flatwoods and savannas, and adults have been observed in crayfish burrows. Juveniles may be restricted to the vicinity of the breeding site in dry years.

Breeding populations of Reticulated Flatwoods Salamanders are vulnerable to conversion of their habitat to pine plantations and by fire-suppression that leads to the loss of wetland vegetation in favor of dense hardwood shrub thickets.

Reticulated Flatwoods Salamanders are listed as Threatened by the U.S. Fish and Wildlife Service.

ORIGINAL ACCOUNT by John G. Palis and
D. Bruce Means; photograph by Dino Ferri.

Reticulated Flatwoods Salamander

Ambystoma californiense Gray, 1853

California Tiger Salamander

California Tiger Salamanders are endemic to California, with their range centered in the eastern Central Valley and with many disjunct populations west of the Central Valley from Clear Lake south to near Santa Barbara.

California Tiger Salamanders breed in fishless, seasonal, and semipermanent wetlands. Newly metamorphosed animals seek cover in ground squirrel burrows and soil cracks. Adult California Tiger Salamanders clearly rely on rodent burrows for underground retreats. Except for the brief breeding season, they spend the entire year in or near these retreats. California Tiger Salamanders commonly settle in burrows in open grassland areas or beneath large oaks; less frequently, they occupy burrows in woodland areas.

Habitat destruction and introduced exotic predators are widely considered to be the primary causes of the decline of California Tiger Salamanders. They also are affected by introduced Western Tiger Salamanders, which hybridize with them in central California.

California Tiger Salamanders are considered a Species of Special Concern across their range by the State of California and are a candidate for listing under the U.S. Endangered Species Act. However, populations near Santa Barbara and Santa Rosa have been listed as Endangered under the U.S. Endangered Species Act.

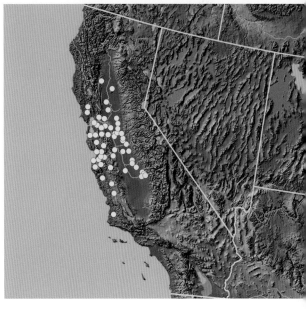

California Tiger Salamander

ORIGINAL ACCOUNT by H. Bradley Shaffer and Peter C. Trenham; photograph by Dante Fenolio.

Ambystoma cingulatum Cope, 1867 (1868)

Frosted Flatwoods Salamander

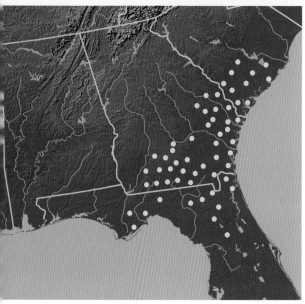

Frosted Flatwoods Salamander

Frosted Flatwoods Salamanders occur on the southeastern Coastal Plain of the U.S. from southern South Carolina through southeastern Georgia into the neck of the Florida peninsula.

Frosted Flatwoods Salamanders tend to live in longleaf pine flatwoods and savannas. They are autumn breeders, and most individuals migrate to breeding sites in October and November and leave again in December through January. Adult Frosted Flatwoods Salamanders typically move during rains associated with passing cold fronts, but will also migrate on rainless nights when soils are near saturation. Breeding sites include isolated swamps where pond cypress or black gum predominate, marshy pasture ponds, road-side ditches, or small, shallow borrow pits. Frosted Flatwoods Salamanders deposit their eggs on bare soil or under cover objects, and the larvae hatch in spring when the ponds flood with water.

Frosted Flatwoods Salamanders no longer occur at many of their historical locations. They have been extirpated from Alabama and appear to have disappeared from two thirds of previously known breeding sites in Florida. There have also been declines in Georgia and South Carolina.

Frosted Flatwoods Salamanders are listed as Threatened by the U.S. government, Protected in Georgia, and State Endangered in South Carolina.

ORIGINAL ACCOUNT by John G. Palis and D. Bruce Means; photograph by Mike Redmer.

Ambystoma gracile (Baird, 1859)

Northwestern Salamander

Northwestern Salamanders are found from northern California north to southeastern Alaska, including Vancouver Island but not the Queen Charlotte Islands, in moist habitats of the Coast Mountains and Cascade Mountains. They occur at elevations from sea level to 3110 m.

Where individual Northwestern Salamanders deposit their eggs depends on whether they have metamorphosed into terrestrial adults or are neotenic and thus have remained in larval form. Metamorphosed females attach egg masses to submerged objects such as tree limbs or branches and cattails from 0.5 to 1 m below the water surface, whereas neotenic adults tend to lay eggs in smaller, looser masses directly on the wetland bottom. Most Northwestern Salamander larvae overwinter and metamorphose in 12 to 14 months, but some may overwinter a second year, and montane populations may overwinter for a third year. Neotenic Northwestern Salamanders may never metamorphose. Terrestrial adult Northwestern Salamanders seem to prefer forest margins and pond banks, hidden underground beneath the leaf litter of the forest or within and under rotten logs, but neotenic adults inhabit the same ponds as do their larvae.

Northwestern Salamanders remain common over many parts of their range. Although, undoubtedly, some populations have disappeared

Northwestern Salamander

due to habitat loss, in other areas ponds constructed for livestock and other uses have probably improved habitat for Northwestern Salamanders in some instances. Logging may temporarily reduce Northwestern Salamander populations but once clear-cuts begin to grow over, there is little correlation between the abundance of salamanders and forest stand age.

Northwestern Salamanders have no conservation status in the U.S. and are considered to be Not at Risk in Canada.

ORIGINAL ACCOUNT by H. Bradley Shaffer; photograph by William Leonard.

Ambystoma jeffersonianum Green, 1827

Jefferson Salamander

Jefferson Salamanders are distributed from eastern Illinois and south central Kentucky northeast to northern Virginia and southwestern New England and into southern Ontario.

Jefferson Salamanders are typically found in undisturbed, well-drained, upland deciduous forests within 200 to 250 m of wetlands suitable for breeding. Adults live in burrows, including rodent burrows, and are often found in well-drained upland forest sites. Jefferson Salamanders are among the first amphibians to emerge from winter dormancy and breed in spring. They typically breed in vernal and semipermanent woodland pools that are not too acidic, but also occasionally in permanent, fishless woodland ponds. Juveniles appear to spend most of their time in underground burrows.

Jefferson Salamanders in the northeastern parts of their range normally coexist with polyploid individuals whose genetic makeup is a mixture of Jefferson Salamander chromosomes and Blue-spotted Salamander chromosomes. Such individuals, all females, usually outnumber the pure, diploid Jefferson Salamanders where they occur. Polyploid individuals that contain two or more Jefferson Salamander genomes are often indistinguishable from Jefferson Salamanders and invariably indicate the presence of a pure Jefferson Salamander breeding population.

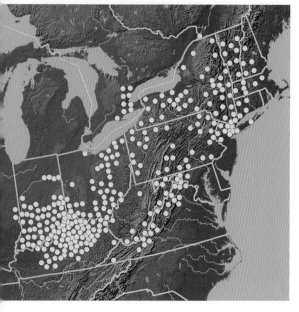

Jefferson Salamander

The presently isolated populations of Jefferson Salamanders in Ontario are remnants of what was once a more extensive and continuous range. The main threats to them are habitat destruction and acidification of breeding ponds.

Jefferson Salamanders are listed as Threatened in Canada federally and in Ontario provincially.

ORIGINAL ACCOUNT by Robert Brodman; photograph by Mike Redmer.

Ambystoma laterale Hallowell, 1856

Blue-spotted Salamander

Blue-spotted Salamanders are found across southern Canada and the northern U.S. from eastern Manitoba, western Minnesota, and Iowa to the Gulf of Saint Lawrence and northern New Jersey and as far north as the La Grande River in Québec.. Isolated populations also exist in Manitoba, Iowa, Indiana, New Jersey, and Nova Scotia.

Blue-spotted Salamanders are most abundant in swamp white oak flatwoods, wooded moraines, moist woodlands with sandy soils, maple–basswood forest, and coniferous forest. They are fossorial and will hide under rocks, stumps, or logs. Blue-spotted Salamanders breed in temporary woodland ponds. Larvae tend to hide among leaf litter and aquatic vegetation, especially where larvae of larger-species salamanders are present.

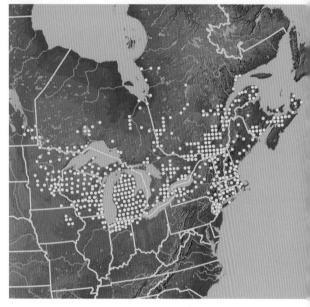

Blue-spotted Salamander

In the southern parts of their range, Blue-spotted Salamanders coexist with all-female populations of polyploid individuals whose genetic makeup is a mixture of Blue-spotted Salamander chromosomes and chromosomes of other, related species, especially Jefferson Salamanders. Such polyploid individuals containing two or more Blue-spotted Salamander genomes can be nearly identical to pure, diploid Blue-spotted Salamanders but usually outnumber them.

Blue-spotted Salamanders are relatively common in many areas but have declined in southern regions of their range and where the upland forest sites and nonacidified, temporary woodland ponds they require have disappeared. Blue-spotted Salamanders are sensitive to forestry management and agriculture practices, and the main threats to them are habitat destruction, land use, and acidification of breeding ponds.

Blue-spotted Salamanders are considered Endangered in New Jersey, Iowa, and Ohio, Threatened in Connecticut, and of Special Concern in Indiana, Vermont, and Massachusetts.

ORIGINAL ACCOUNT by Robert Brodman; photograph by Patrick Moldowan.

Ambystoma mabeei Bishop, 1928

Mabee's Salamander

Mabee's Salamanders occur entirely in the Atlantic Coastal Plain from the Painkatank River in southeast Virginia to the Savannah River at the southern tip of South Carolina.

Adult Mabee's Salamanders breed in small, shallow, typically ephemeral to semipermanent wetlands that are usually free of fishes. A wide variety of pools support breeding populations, including farm ponds, water-filled foxholes, vernal pools in pine and hardwood forests, Carolina bays, sinkhole ponds, and cypress–tupelo ponds in pinewoods. Mabee's Salamanders use terrestrial habitats extensively outside of the breeding period, including open fields, pine forest, and hardwood forest.

Threats to Mabee's Salamanders include habitat fragmentation, aquatic and terrestrial habitat loss, road mortality, and alteration of hydrology, mostly due to urbanization.

Mabee's Salamanders are listed as Threatened in Virginia, which has dedicated a portion of the Grafton area sinkhole pond complex that contains a Mabee's Salamander site as a Virginia Department of Conservation and Recreation Natural Area Preserve.

ORIGINAL ACCOUNT by Joseph C. Mitchell; photograph by Pierson Hill.

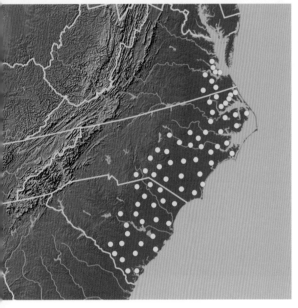

Mabee's Salamander

Ambystoma macrodactylum Baird, 1849[11]

Long-toed Salamander

Long-toed Salamanders occur from the Alaskan Peninsula across British Columbia, south through Washington into Oregon and the Sierra Nevada of California, and across the Rocky Mountains into eastern Alberta, western Montana, and central Idaho. Isolated populations occur in coastal California around Monterey Bay.

Long-toed Salamanders breed in a variety of habitats, including seeps, backwaters of slow-flowing streams, lower-elevation temporary pools, and small to large permanent lakes and ponds at higher elevations and higher latitudes. They exhibit strong breeding-site fidelity and generally will remain within 100 m of breeding ponds; however, longer-distance movements may occur. Terrestrial movements by postmetamorphic individuals are generally associated with rains, high soil moisture, and air temperatures above freezing. Adult Long-toed Salamanders are fossorial and occur in a wide range of habitats, including semiarid grasslands and sagebrush steppes, alpine meadows, dry oak woodlands, humid coniferous forests, rocky shorelines of subalpine lakes, beaver ponds, disturbed agricultural areas, timber harvest areas, pastures, and suburban forests.

Limiting factors for Long-toed Salamanders include habitat alteration, destruction of critical habitats, and the presence of introduced, nonnative, predatory fish.

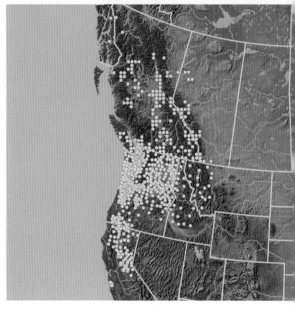

Long-toed Salamander

Long-toed Salamanders in the Monterey Bay area of California at the southern edge of it range are federally listed in the U.S. as Endangered. Long-toed Salamanders are a species of Special Concern in Alberta. In Alaska, Long-toed Salamanders are ranked as "Imperiled" by the Alaska State Heritage system.

ORIGINAL ACCOUNT by David S. Pilliod and Julie A. Fronzuto; photograph by William Leonard.

Ambystoma maculatum (Shaw, 1802)

Spotted Salamander

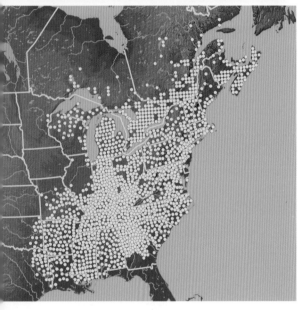

Spotted Salamander

Spotted Salamanders are distributed throughout the eastern U.S. and Canada from the Maritime Provinces, southern Québec, and the northern coasts of the Great Lakes in Ontario south to the Gulf Coastal Plain in eastern Texas, Louisiana, and Mississippi, except for most of Florida and parts of southern Atlantic Coastal Plain.

Spotted Salamanders primarily inhabit hardwood and mixed coniferous–deciduous forest. They are often found in lowland forests and are occasionally found in relatively open habitats such as meadows, though usually near forest edges. Along the Atlantic Coastal Plain, Spotted Salamanders are typically found in hardwood, bottomland habitats with low soil temperatures and high soil moisture. Although they are not typically a montane species, Spotted Salamanders can occur at high elevations in the Appalachian Mountains when suitable breeding sites are available. Both juvenile and adult Spotted Salamanders live under cover objects and in burrows created by shrews, moles, and other small mammals.

Adult Spotted Salamanders migrate in early spring from their terrestrial overwintering sites into seasonally available, ephemeral, fishless wetlands to breed in roadside ditches, tire ruts in dirt roads, artificial ponds, floodplain wetlands, and marshes. Newly metamorphosed Spotted Salamanders that are dispersing from breeding ponds will hide beneath rocks and logs near the pond margin.

Spotted Salamanders remain widespread and abundant throughout their range, though they are vulnerable to development that alters or eliminates habitats.

ORIGINAL ACCOUNT by Wesley K. Savage and Kelly R. Zamudio; photograph by Mike Redmer.

Ambystoma mavortium

Western Tiger Salamander

Western Tiger Salamanders occur throughout the interior North American plains region and eastern Rocky Mountains from the foothills and prairies of Alberta, Saskatchewan, and Manitoba south to Arizona, New Mexico, and Texas and on into northern Mexico (not shown). Western Tiger Salamanders are also found in eastern Washington and adjacent Oregon, Idaho and British Columbia. Scattered, introduced populations are present throughout the West.

Western Tiger Salamanders can either transform and be fully terrestrial as adults or else remain aquatic as gilled, neotenic individuals. Terrestrial adults require deep, friable soils, where they either dig their own burrows or use burrows made by ground squirrels or pocket gophers. Breeding habitats for Western Tiger Salamanders consist of seasonal, semipermanent, and permanent fishless wetlands, including roadside ditches, quarry ponds, cattle tanks, subalpine lakes, and sluggish streams. In some populations, newly metamorphosed Western Tiger Salamanders remain near wetlands, while in others they will migrate some distance away, depending upon terrestrial conditions.

The biggest threat to existing Western Tiger Salamander populations comes from wetland destruction and alteration, though they are fairly tolerant of agriculture. Larvae and neotenic adult Western Tiger Salamanders are threatened by droughts and by introduced fishes that will prey on them.

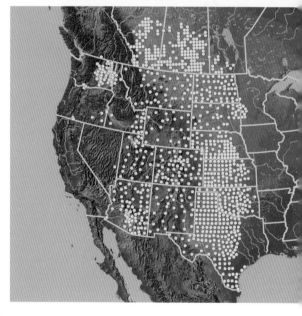

Western Tiger Salamander

Western Tiger Salamanders are listed as Protected in Arizona. In Canada, they are on the British Columbia Red List of Threatened and Endangered Species and are listed as Endangered in British Columbia and as of Special Concern in Alberta, Saskatchewan, and Manitoba under the federal Species at Risk Act.

ORIGINAL ACCOUNT by Michael J. Lannoo; photograph by Mike Redmer.

Ambystoma opacum (Gravenhorst, 1807)

Marbled Salamander

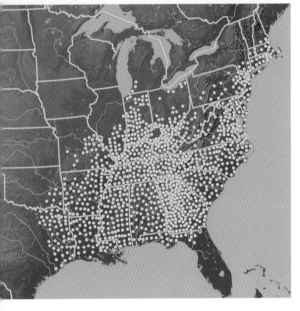

Marbled Salamander

Marbled Salamanders range throughout much of the eastern U.S. from eastern Texas and Oklahoma, northeast through Illinois and Indiana to southern New Hampshire and central Massachusetts, and south to north Florida. Disjunct populations also occur in various places in the north of the range.

Marbled Salamanders breed during the autumn on land, and females remain with their eggs for variable lengths of time until spring run-off floods the fish-free vernal wetlands needed by their larvae. These habitats include upland hardwood swamp forests, quarries, vernal ponds, Carolina bays, and floodplain pools. Adult Marbled Salamanders inhabit mature deciduous forests, mixed hardwood and pine stands in both floodplains and uplands, where they reside within the forest under leaf litter and in small mammal burrows.

Because Marbled Salamanders rely on the presence of small, isolated, seasonal wetlands and intact forested floodplain habitats, their abundance has declined as such wetland habitats and surrounding forests have been destroyed.

Marbled Salamanders are listed as Threatened in Massachusetts and Michigan and as Protected in New Jersey.

ORIGINAL ACCOUNT by David E. Scott; photograph by Brad Moon.

Ambystoma talpoideum (Holbrook, 1838)

Mole Salamander

Mole Salamanders are distributed along the southern Atlantic and Gulf Coastal Plains from the Pee Dee River in central South Carolina to the Naches River in eastern Texas, and north along the Mississippi River Valley to southern Illinois. Disjunct populations are scattered in Kentucky, Virginia, Tennessee, North Carolina, northern South Carolina, northern Georgia, and northern Alabama.

Mole Salamanders breed in forested, fishless wetlands that range from seasonal, semipermanently to permanently water-filled sites, including Carolina bays, gravel pits, and roadside ditches. Adult Mole Salamanders may be metamorphosed terrestrial individuals or remain as aquatic gilled neotenic individuals. Populations associated with seasonal and/or semipermanent wetlands produce terrestrial adults, whereas populations associated with permanent wetlands produce neotenic adults. In expansive lowland floodplain forests, terrestrial adult Mole Salamanders are found in areas near gum and cypress ponds, whereas in upland areas they inhabit mixed conifer–hardwood forests.

The conversion of moist forests to agricultural use or given over to urban or suburban development can lead to the loss of many populations of Mole Salamanders. Clear-cutting of forests can also reduce their abundance.

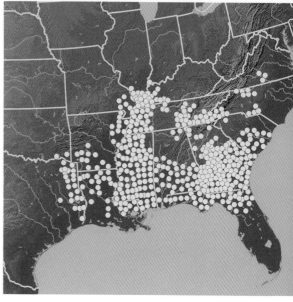

Mole Salamander

Mole Salamanders are listed as a species of Special Concern in North Carolina and a species In Need of Management in Tennessee.

ORIGINAL ACCOUNT by Stanley E. Trauth; photograph by Mike Redmer.

Ambystoma texanum (Matthes, 1855)

Small-mouthed Salamander

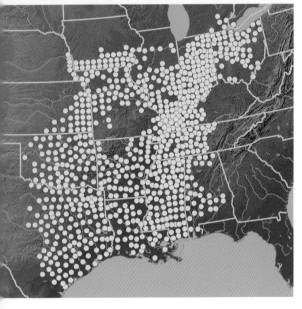

Small-mouthed Salamander

Small-mouthed Salamanders are distributed throughout the central U.S., except for the Ozark Plateau, from southeastern Nebraska, eastern Kansas, Oklahoma, and eastern Texas to southeastern Michigan and Ohio south through western Kentucky, Tennessee, and Alabama to the Gulf Coast from Mobile Bay to near the Nueces River. Small-mouthed Salamanders are also found on Pelee Island in Ontario and the Bass Islands of western Lake Erie.

Small-mouthed Salamanders breed in fishless seasonal and semipermanent wetlands, including prairie potholes, forested wetlands, oxbows, ditches, borrow pits, flooded fields, and occasionally stream pools. Adult Small-mouthed Salamanders will reside beneath logs or other cover objects or in burrows, including crayfish burrows, in humid forest floors near the margins of their breeding wetlands. They surface during rainy nights.

Many populations of Small-mouthed Salamanders have probably been lost because of wetland drainage and deforestation due to agricultural practices and urban or suburban development.

Small-mouthed Salamanders are listed as Endangered by the State of Michigan, and legally protected under Michigan's Natural Resources and Environmental Protection Act. In Canada, Small-mouthed Salamanders are listed as Endangered under the federal Species at Risk Act and Ontario's Endangered Species Act.

ORIGINAL ACCOUNT by Stanley E. Trauth; photograph by Bill Peterman.

Ambystoma tigrinum (Green, 1825)

Eastern Tiger Salamander

Eastern Tiger Salamanders occur in scattered locations along the Gulf Coastal Plain from east Texas to northern Florida, east of the Appalachians up the Atlantic Coastal Plain to Long Island in New York, and north to the Great Lakes west of the Appalachians to enter, barely, into southern Ontario and southeastern Manitoba. They were historically found throughout western Ohio, the Lower Peninsula of Michigan, Indiana, Illinois, Iowa, Tennessee, and much of Wisconsin, Minnesota, Missouri, and western Kentucky.

Eastern Tiger Salamanders tend to be associated with grasslands, savannas, woodland edges, and, less so, forests. Adult Eastern Tiger Salamanders burrow and require deep, friable soils. They breed in seasonal, semipermanent and permanent wetlands, including roadside ditches, quarry ponds, cattle tanks, and sluggish streams. Eastern Tiger Salamander larvae generally can survive only in wetlands without predatory fish and where they assume the role of top aquatic carnivores.

Eastern Tiger Salamanders are tolerant of agriculture and are usually the most common salamanders throughout the Midwest. The biggest threats to populations come from continued wetland destruction and wetland alteration. Introduced fishes will reduce or eliminate Eastern Tiger Salamander populations. Deforestation and acid deposition are also problems. The losses

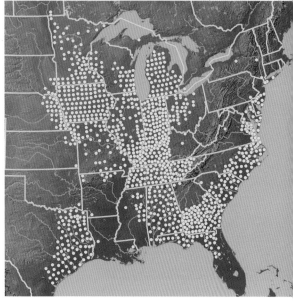

Eastern Tiger Salamander

of populations of Eastern Tiger Salamanders have resulted in a pattern of range fragmentation, with sporadic, anthropogenically assisted invasions of new habitats.

Eastern Tiger Salamanders are listed as Endangered in Delaware, New York, New Jersey, and Maryland and of Special Concern in North Carolina and South Carolina. In Ontario, they are listed as Extirpated from their only known historical locality in the province, Point Pelee.

ORIGINAL ACCOUNT by Michael J. Lannoo; photograph by Mike Redmer.

Ambystoma sp.

Unisexual Jefferson Salamander *complex*

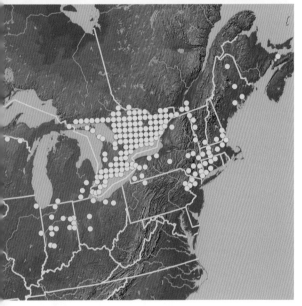

Jefferson Salamander complex

Unisexual populations of female salamanders of the Jefferson Salamander *complex* are sporadically distributed across the lower and eastern Great Lakes region, New York, Vermont, New Hampshire, southern Quebec, southern New England, and northern New Jersey, with isolated populations in Wisconsin, Maine, northern New Brunswick and central Nova Scotia.

Jefferson Salamander *complex* populations always coexist with populations of either Jefferson Salamanders or Blue-spotted Salamanders, and their relationship to these and several other related species of salamanders is complicated and bizarre[12]. They are intermediate in morphology and behavior between Jefferson Salamanders and Blue-spotted Salamanders, are almost invariably polyploid, and are only of one sex. Yet salamanders of the Jefferson Salamander *complex* are not hybrids in any usual sense of the word. Indeed, there is no evidence that they are, or ever were, the result of interbreeding between any two of the diploid species with which they are associated.

Instead, Jefferson Salamander *complex* individuals practice an unusual reproductive strategy. The eggs they lay usually do not have a reduction in the number of sets of chromosomes they possess, as would be normal, and a sperm from a diploid male of another species is required to initiate egg development. The sperm's genome may or may not be incorporated into the genetic makeup of the developing zygote. In effect, they steal sperm from males of other species.

The result of this strange reproductive system is that salamanders of the Jefferson Salamander *complex* are usually triploid, but may be diploid, tetraploid, or even pentaploid. They may also incorporate nuclear genomes of at least five distinct species of salamanders, including Small-mouthed Salamanders, Eastern Tiger Salamanders, and Streamside Salamanders, as well as Jefferson Salamanders and Blue-spotted Salamanders. However, all Jefferson Salamander *complex* individuals carry at least one genome like that of Blue-spotted Salamanders and have mitochondria like those of Streamside Salamanders.

Jefferson Salamander *complex* adults are found in upland forests as well as bottomland forests, often associated with sandy soils. They may remain near the surface most of the year until late autumn. They breed in fishless ponds in a variety of wooded and semiwooded habitats including ponds, wetlands, ditches, and sloughs.

Polyploid, unisexual lineages of these salamanders are recognized as animals of Special Concern in Connecticut and as Endangered in Illinois and New Jersey.

ORIGINAL ACCOUNT by Christopher A. Phillips and Jennifer Mui; photograph by Michael Graziano.

Amphiuma means Garden, 1821

Two-toed Amphiuma

The range of Two-toed Amphiumas includes the Gulf and Atlantic Coastal Plains from about New Orleans to southeastern Virginia and all of Florida except for the Florida Keys.

Two-toed Amphiumas occupy a great variety of aquatic habitats, including permanent ponds and lakes, preferring relatively shallow, heavily vegetated habitats. They also inhabit isolated, ephemeral wetlands, wet prairies and marshes, swamps, and the Florida Everglades. They are common in canals and drainage ditches, preferring to burrow in mucky substrates. They are often found inhabiting crayfish burrows. Though otherwise completely aquatic, female Two-toed Amphiumas deposit their eggs in moist, terrestrial sites.

Two-toed Amphiumas are considered common, but destruction and degradation of wetlands has likely caused the loss or decline of many local populations.

Two-toed Amphiumas have no federal protection in the U.S. and are not listed at any level in the eight states where they occur.

ORIGINAL ACCOUNT by Steve A. Johnson and Richard B. Owen; photograph by Pierson Hill.

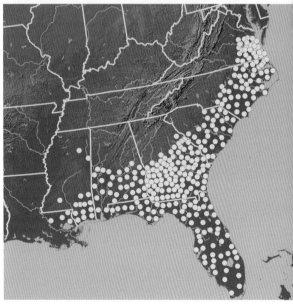

Two-toed Amphiuma

Amphiuma pholeter Neill, 1964

One-toed Amphiuma

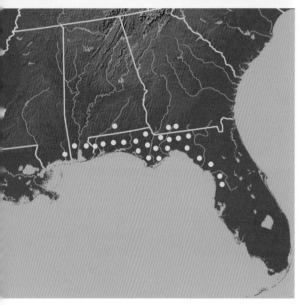

One-toed Amphiuma

One-toed Amphiumas are found in a narrow band along the eastern Gulf Coast Plain of the southeastern U.S. from the Pascagoula River in southern Mississippi to the Chassahowitzka Swamp in northern Florida, ranging only up to about 80 to 120 km inland from the seashore.

One-toed Amphiumas are primarily found in deep, liquid, amorphous muck in slow-moving streams, swampy and periodically inundated floodplains, and mixed hardwood and cypress bottomlands. They are rarely found in shallow muck deposits, presumably because that would increase their vulnerability to predators. The muck derived from hardwood and cypress litter, as compared with peat, is usually so far decomposed that it no longer features any large pieces of wood or leaf litter. One-toed Amphiumas cannot move through fibrous peat.

In winter, One-toed Amphiumas are occasionally found under large logs buried along stream courses in first-order stream valleys, suggesting that some individuals move upstream to find protection from cold weather in warm seeps.

One-toed Amphiumas are considered Rare in Florida and Georgia and are under consideration for Endangered status in Mississippi.

ORIGINAL ACCOUNT by D. Bruce Means; photograph by Michael Graziano.

144   Family Amphiumidae

Amphiuma tridactylum Cuvier, 1827

Three-toed Amphiuma

Three-toed Amphiumas occur in the Gulf Coast Plain from the Brazos River Valley in Texas to central Alabama and north up the lower Mississippi River to southeastern Missouri and extreme southwestern Kentucky.

Three-toed Amphiumas occupy bottomland swamps, bayous, cypress swamps, and streams. Like the other species of amphiumas, Three-toed Amphiumas can co-exist with fishes in these water bodies. They are especially abundant in drainage ditches in suburban and agricultural areas of the lower Mississippi River. Three-toed Amphiumas will frequently inhabit crayfish burrows. Though almost wholly aquatic, Three-toed Amphiumas will move overland during and following heavy rains as far as 12 m away from the water's edge.

Three-toed Amphiumas are considered widespread and abundant. They are subject to a modest harvest in Louisiana, usually as bycatch during crayfish trapping, and they are sold as pets or to biologic suppliers for education and research.

ORIGINAL ACCOUNT by Jeff Boundy; photograph by Paul Crump.

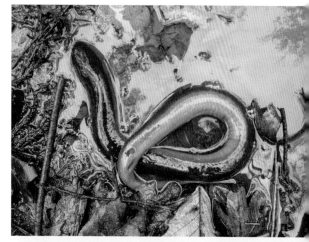

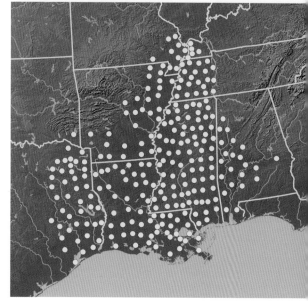

Three-toed Amphiuma

Cryptobranchus alleganiensis (Daudin, 1803)

Hellbender

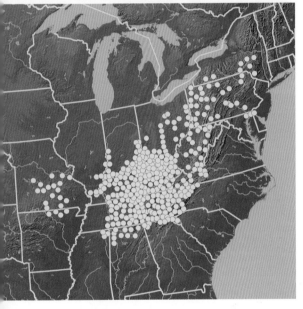

Hellbender

Hellbenders are native to river systems in upland areas in the eastern U.S. from southern New York to northeastern Mississippi and northern Alabama, including the Susquehanna River, tributaries of the Savannah River, the Tennessee River, and the Ohio River. They also inhabit the Missouri River drainage, the Meramec River, and the White River in portions of Missouri and northern Arkansas.

Hellbenders are found in fast-flowing streams containing abundant cover in the form of large flat rocks, bedrock shelves and crevices, and logs.

The range of Hellbenders has shrunk considerably as a result of human modification of their stream habitats, including industrialization, channelization, and impoundment of streams, that has resulted in increased siltation, contamination from mine wastes and agricultural runoff, and thermal pollution.

Hellbenders are classified as Endangered in Illinois, Indiana, Maryland, and Ohio; as Rare in Georgia; and of Special Concern or as Species of Concern in New York, North Carolina, and Virginia. They are on the Watch List in Missouri and are deemed in Need of Management in Tennessee. Other states, such as Alabama, Arkansas, Kentucky, Mississippi, South Carolina, and West Virginia track hellbender distribution records in a database, but do not generally afford them protection from take.

ORIGINAL ACCOUNT by Christopher A. Phillips and W. Jeffrey Humphries; photograph by William Leonard.

Dicamptodon aterrimus (Cope, 1867)

Idaho Giant Salamander

Idaho Giant Salamanders inhabit north central Idaho from the Coeur d'Alene River south to the Salmon River and several headwater streams south of Saltese and Deborgia in extreme western Montana.

Idaho Giant Salamanders can be locally abundant in or near headwater streams in coniferous forest watersheds. Neoteny, whereby sexually mature individuals retain their larval form throughout life, is common among Idaho Giant Salamanders. Terrestrial adults are rarely encountered, and knowledge of their habits is scarce, but when they are encountered they are usually under logs and bark. Courtship among Idaho Giant Salamanders likely takes place in hidden nest chambers beneath logs, stones, and crevices in small streams. The larval period can last 3 years, and the animals that do metamorphose move from streams into nearby humid coniferous forest floors.

Idaho Giant Salamanders have likely been reduced in abundance in heavily logged watersheds, where their larvae may be adversely affected by sedimentation and increases in water temperature.

Idaho Giant Salamanders are a Species of Concern in Montana.

ORIGINAL ACCOUNT by Kirk Lohman and R. Bruce Bury; photograph by William Leonard.

Idaho Giant Salamander

Dicamptodon copei Nussbaum, 1970

Cope's Giant Salamander

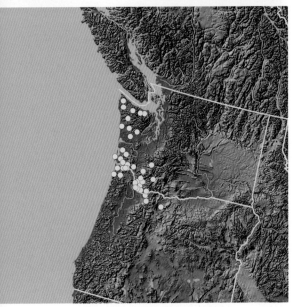

Cope's Giant Salamander

Cope's Giant Salamanders occur from the Cascades Mountains south of Mt. Rainier, the Willapa Hills, and the Olympic Peninsula and from the vicinity of the Columbia River Gorge in northwestern Oregon.

Cope's Giant Salamanders appear to be nearly obligatorily neotenic as adults, as few metamorphosed individuals have ever been found. Larvae and neotenic adult Cope's Giant Salamanders are usually associated with pools in small to moderately sized rocky mountain streams and, occasionally, montane lakes. They often use large stones for cover.

There has been no apparent reduction in the range of Cope's Giant Salamanders. They are widely distributed through much of their range and can be locally abundant. While little is known about the effects of timber harvesting on Cope's Giant Salamanders, it is likely that development and deforestation may have resulted in the reduction of some populations.

Cope's Giant Salamanders are listed as Protected in Oregon.

ORIGINAL ACCOUNT by Lawrence L.C. Jones and R. Bruce Bury; photograph by William Leonard.

Dicamptodon ensatus Eschscholtz, 1833

California Giant Salamander

California Giant Salamanders occur in the central Pacific Coastal region of California from around Monterey Bay north to Anchor Bay.

California Giant Salamanders are associated with permanent and semipermanent streams and breed in subterranean crevices, fissures, and crannies in running water. Younger larvae are found in slowly moving streams near the shoreline, while older larvae tend to be found in the main stream channel. California Giant Salamander larvae often hide under gravel, cobbles, or other objects in streams. Neotenic individuals that retain their external gills as adults are present in many populations. Terrestrial adults occasionally are found under cover objects, such as rocks and logs, near to streams, or under stones in streams during the breeding season.

Populations of California Giant Salamanders have undoubtedly been lost because of the intense habitat alterations characteristic of many parts of their range. Because they have a small range that exists within an area of intense human activity, they are threatened by siltation of their stream habitats and urbanization.

California Giant Salamanders are not listed under any state or federal laws or regulations in the U.S.

ORIGINAL ACCOUNT by R. Bruce Bury; photograph by William Leonard.

California Giant Salamander

Dicamptodon tenebrosus (Baird and Girard, 1852)

Coastal Giant Salamander

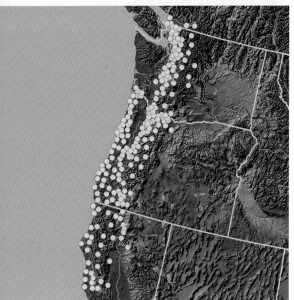

Coastal Giant Salamander

Coastal Giant Salamanders occur from near Anchor Bay in California north to the Chilliwack River basin in extreme southwestern British Columbia along the Coast and Cascades Ranges, except the Olympic Peninsula.

Coastal Giant Salamanders are found in clear-running streams without fine sediments in unlogged mature and old-growth forests. Neotenic individuals, which retain larval form as adults, typically inhabit large perennial streams, rivers, lakes, and ponds. Smaller streams that may dry up predominantly harbor populations of individuals that metamorphose into terrestrial adults. Newly metamorphosed Coastal Giant Salamanders may venture into upland habitats some 400 m or more from water during rainy periods. Terrestrial adults prefer forested habitats, where they are found under rocks and in logs, root channels, and burrows. Neotenic individuals require permanent water, where they seek cover in coarse substrates.

There has been some fragmentation of the range of Coastal Giant Salamanders as a result of habitat alterations, mostly due to forestry practices. Increased siltation negatively affects Coastal Giant Salamanders by filling in the rocky interstices under cover objects, particularly in low-gradient streams that do not adequately flush sediments. But despite the negative effects of habitat alteration, Coastal Giant Salamanders seem to be more resilient than other stream-associated amphibians in much of their range.

In British Columbia, where they have a limited distribution, Coastal Giant Salamanders are listed as Threatened federally and are on the Red List of Threatened and Endangered Species.

ORIGINAL ACCOUNT by Lawrence L.C. Jones and Hartwell H. Welsh Jr.; photograph by Brad Moon.

Aneides aeneus Cope and Packard, 1881

Green Salamander

Green Salamanders range along the Appalachian Mountains and Cumberland Mountains from northern West Virginia and southeastern Pennsylvania south and west to northern Alabama and the extreme northeast of Mississippi. They also occur in the Blue Ridge region of southwestern North Carolina, northwestern South Carolina, and northeastern Georgia. Disjunct populations have been found in southern Indiana and parts of Tennessee. Green Salamanders occur at elevations up to 1341 m on Cold Mountain in North Carolina.

A favored microhabitat of Green Salamanders is beneath the bark of downed trees and logs. For a time in the 1930s, Green Salamanders reached tremendous population numbers under the bark of millions of dead American chestnut trees. It is likely that the large, thick slabs of bark on large old-growth logs provided Green Salamanders with more favorable conditions for foraging and nesting than the smaller logs found in secondary forests today.

Green Salamanders are sensitive to the removal of rocks and trees, which eliminates habitat, and road cuts or other, larger, corridors adjacent to emergent rocks and outcrops that can result in increasing temperatures and decreasing moisture levels.

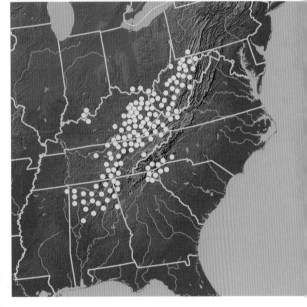

Green Salamander

Green Salamanders are listed as Endangered in Indiana, Ohio, Maryland, and Mississippi, Threatened in Pennsylvania, Protected in Georgia, Rare in North Carolina, and of Special Concern in South Carolina and West Virginia.

ORIGINAL ACCOUNT by Thomas K. Pauley and Mark B. Watson; photograph by Mike Redmer.

Aneides ferreus Cope, 1869

Clouded Salamander

Clouded Salamander

Clouded Salamanders are found in coastal Oregon and extreme northern California from the Columbia River south through the Siskiyou, Coast, and western Cascades mountains of Oregon and just into northwestern California. Clouded Salamanders occur at elevations up to 1525 m.

Clouded Salamanders occur in high densities in old-growth forests and in recently cut or burned areas in association with stumps, decaying logs, and coarse, woody debris, preferring more humid forest stands that are most commonly found in decaying logs and stumps with intact bark. Juvenile Clouded Salamanders prefer bark litter over rock or leaf litter, whereas adults are found associated with stumps and logs in Douglas fir forests, crevices in rock outcrops and road cuts, and talus.

Populations of Clouded Salamanders have certainly been lost because of forestry-management practices and urban sprawl. It is doubtful that this species survives in areas where forests are intensively managed on short rotation cycles because of severe reduction in moisture conditions and the amount of large woody debris.

Clouded Salamanders are listed as Protected in Oregon.

ORIGINAL ACCOUNT by Nancy L. Staub and David B. Wake; photograph by Mike Redmer.

Aneides flavipunctatus (Strauch, 1870)

Black Salamander

Black Salamanders occur in northern California and extreme southern Oregon as well as the Santa Cruz Mountains, separated by a large gap from northern populations. Black Salamanders generally are found at elevations below 600 m, but occur as high as 1700 m.

Black Salamanders are primarily ground dwellers associated with lowland forests, grassy meadows, pastures, burned areas, talus slopes, and streamside habitats. They live under rocks and logs or in wet soil along streams in areas that receive over 75 cm of precipitation annually. Black Salamanders are active year-round, though they will move underground during the dry season. Females probably lay eggs in July or early August in cavities as much as 38 cm below ground.

Black Salamanders were once considered common in many areas of their range but have become rare in recent years. The proliferation of vineyards in northern California has destroyed much of the prime habitat for Black Salamanders.

Black Salamanders are listed as Protected in Oregon.

ORIGINAL ACCOUNT by Nancy L. Staub and David B. Wake; photograph by Todd Pierson.

Black Salamander

Aneides hardii (Taylor, 1941)

Sacramento Mountain Salamander

Sacramento Mountain Salamander

ORIGINAL ACCOUNT by Cindy Ramotnik; photograph by Gary Nafis.

Sacramento Mountain Salamanders are found only in the Capitan, White, and Sacramento mountains of south central New Mexico above elevations of 2400 m.

In mixed spruce fir forest, Sacramento Mountain Salamanders can be found within or under coniferous logs in advanced stages of decay, under bark, or in small cracks and chambers near the inner bark surface in less decayed logs. Above the timberline they are associated with rocks and mats of mosses and lichens. During periods of drought, Sacramento Mountain Salamanders retreat beneath large decayed logs or other surface objects, or into subterranean retreats. They respond to periods of decreased temperature and precipitation by reducing their surface activity.

Sacramento Mountain Salamanders are potentially vulnerable to actions such as logging and fire that dry the habitat. However, these salamanders have endured bouts of sometimes intense logging activity in the Sacramento Mountains over the past 60 to 90 years, and there is no evidence that populations have been eliminated. Sacramento Mountain Salamanders apparently survive well the frequent, low-intensity fires that occurred historically in the Sacramento Mountains. The salamanders minimize moisture loss on burned sites by aggregating in large numbers beneath logs.

Sacramento Mountain Salamanders are listed as Threatened in New Mexico.

Aneides lugubris (Hallowell, 1949)

Arboreal Salamander

Arboreal Salamanders occur along the Pacific Coast from northern California south into northern Baja California in Mexico (not shown). Their range also includes South Farallon, Santa Catalina, Los Coronados, and Año Neuvo Islands off the coast of California and a geographically isolated cluster of populations in the foothills of the Sierra Nevada

Arboreal Salamanders inhabit a variety of terrestrial and arboreal habitats in coastal oak woodlands and black oak and yellow pine forests. They are found under rocks and woody surface cover, in decaying stumps and logs, in decay holes in trees, and in rock crevices. Individuals have been found over 18 m above ground in trees. They are nocturnal and feed most actively under moist or wet conditions. Arboreal Salamanders are more tolerant of dry conditions than are many species of salamanders and are often among the last salamanders to retreat underground or into tree holes to avoid desiccation. Arboreal Salamanders commonly lay their eggs in decay holes of live oak trees up to 9 m above ground, or else under rocks set deeply in the ground or in logs or under surface cover objects. Females coil around their eggs and males are often in attendance as well. Arboreal salamanders are well known for their aggressive tendencies and strong jaws.

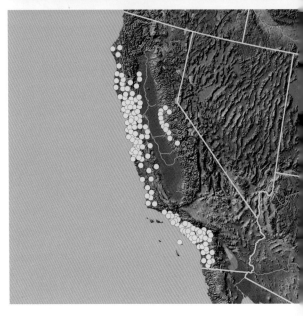

Arboreal Salamander

Populations of Arboreal Salamanders have been eliminated as coastal California habitats have been developed, but the species survives in many urbanized regions where adequate cover is present. The decline of live oaks in California has likely had negative effects on Arboreal Salamander populations.

Arboreal Salamanders have no conservation status at either the state or federal level in the U.S.

ORIGINAL ACCOUNT by Nancy L. Staub and David B. Wake; photograph by William Leonard.

Aneides vagrans Wake and Jackman, 1998

Wandering Salamander

Wandering Salamanders are found in the coastal forests of northern California and British Columbia, but in neither Oregon nor Washington. Their distribution in California extends from near the border with Oregon south to the vicinity of Stewart's Point. In British Columbia, Wandering Salamanders are widespread on Vancouver Island and neighboring islands and occur in a small area of mainland British Columbia. Genetic and historical evidence suggests that Wandering Salamanders may have been introduced to British Columbia from California in shipments of tan oak bark, which in the past was used extensively in the tanning of leather.

Adult Wandering Salamanders are found associated with stumps and logs in Douglas fir forests and under redwood slabs in redwood forests. They have a tendency to be in but not under rotten logs in edge and open habitat such as abandoned pastures and logging platforms in forest clearings and in secondary forests. Juveniles may prefer bark litter over rock or leaf litter. Wandering Salamanders also inhabit the canopy of old-growth redwood forests, where they have been found in humus accumulations in trunk crotches, on limbs, under bark, and in cracked and rotting wood of broken limbs and trunks.

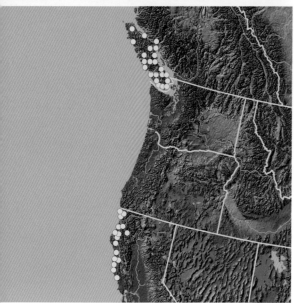

Wandering Salamander

Wandering Salamanders reach their highest densities in old-growth forests. Yet although populations certainly have been lost because of forestry and urban sprawl, they can also be found at high densities in regenerating forest. In British Columbia, forest harvesting does not appear to have had long-term effects on their abundance.

Wandering Salamanders are not listed under any provincial, state, or federal regulations in either Canada or the U.S.

ORIGINAL ACCOUNT by Nancy L. Staub and David B. Wake; photograph by William Leonard.

Batrachoseps altasierrae (Jockusch, Martínez-Solano, Hansen, and Wake 2012)[13]

Greenhorn Mountains Slender Salamander

Greenhorn Mountains Slender Salamanders occur in the southern Sierra Nevada in Kern and Tulare counties, California. They range from the northern Lower Kern River Canyon to the Tule River and Little Kern River drainages, with populations also present on the western margin of the Kern Plateau. Greenhorn Mountains Slender Salamanders are found at relatively high elevations, from 900 to 2440 m above sea level.

Greenhorn Mountains Slender Salamanders are listed currently in California and federally in the U.S. as a Species of Special Concern, even though populations of these salamanders appear to be secure throughout their range. This is because Greenhorn Mountains Slender Salamanders were thought to be the same species as Relictual Slender Salamanders, *B. relictus,* at the time that species was assessed.

ACCOUNT by John Cavagnaro and Michelle S. Koo; photograph by Gary Nafis.

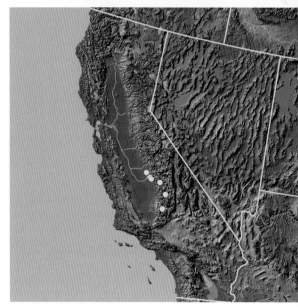

Greenhorn Mountains Slender Salamander

Batrachoseps attenuatus (Eschscholtz, 1833)

California Slender Salamander

California Slender Salamanders occur along the California Coast and adjacent Inner Coast Range from north and east of Monterey Bay northward to extreme southwestern Oregon and in the western foothills of the northern and central Sierra Nevada. There are also populations just north of California's Central Valley in the Little Cow and Clipkapudi Creek drainages.

Along the coast, California Slender Salamanders live at low elevations in coastal redwood forest, humid mixed evergreen forests, or oak woodlands. In the Sierra Nevada, though, California Slender Salamanders are principally associated with pine–oak woodland and chaparral. Individuals occur under logs, bark, rocks, boards, and other surface cover and in damp leaf litter. They are most active at the surface during the rainy season.

California Slender Salamander population densities in the coastal portion of their range remain high, though large portions of their historical range have been modified by development for housing, agriculture, and other activities. However, as modern agriculture replaces small creeks draining into the Central Valley with drainage ditches, California Slender Salamanders, which previously lived in narrow strips of riparian oak woodlands, can survive only by adapting to urban or suburban gardens and vacant lots.

California Slender Salamander

California Slender Salamanders are not listed either federally or at the state level in the U.S.

ORIGINAL ACCOUNT by Robert W. Hansen and David B. Wake; photograph by Dante Fenolio.

Batrachoseps bramei (Jockusch, Martínez-Solano, Hansen, and Wake 2012)[13]

Fairview Slender Salamander

The range of Fairview Slender Salamanders extends only 30 km north to south from the Upper Kern River Canyon and along the western side of Lake Isabella from at least 1 km north of the confluence of South Falls Creek and the Kern River south to Wofford Heights in California.

Fairview Slender Salamanders are associated with talus at the base of slopes and are often found under these rocks, as well as logs, leaf litter, and other cover in chaparral. These habitats experience periodic fires, and while chaparral fires do not harm salamanders, human activities aimed at fire suppression might be detrimental to them.

Fairview Slender Salamanders are relatively abundant throughout their range in California. Besides fire-suppression activities, threats include roadwork associated with Mountain Highway 99. Most known populations are located on public lands associated with Sequoia National Forest and are therefore relatively protected.

Fairview Slender Salamanders are not listed in the U.S. under any state or federal laws or regulations.

ACCOUNT by John Cavagnaro and Michelle S. Koo; photograph by Gary Nafis.

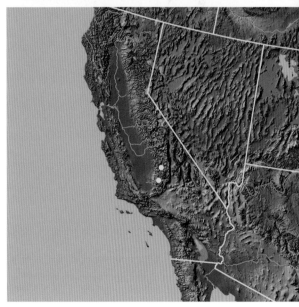

Fairview Slender Salamander

Batrachoseps campi Marlow, Brode, and Wake, 1979

Inyo Mountains Salamander

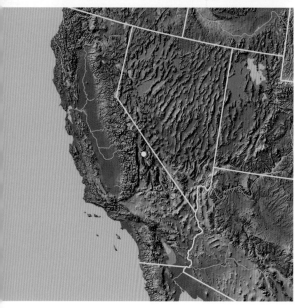

Inyo Mountains Salamander

Inyo Mountains Salamanders occur on the slopes of the Inyo Mountains in California's northern Mojave Desert at elevations from 490 to 2590 m. All the sites inhabited by Inyo Mountains Salamanders are small areas of suitable habitat bordered by large expanses of inhospitable desert or semidesert terrain.

Inyo Mountains Salamanders populations are largely restricted to perennial springs and seepages, especially where solid-rock cliffs, outcrops, or talus are in contact with surface water flow. These areas are grown to willow, wild rose, and coyote brush, which often form such dense thickets that direct sunlight rarely reaches ground level. Inyo Mountains Salamanders have been found under rocks resting on wet substrates, sometimes under woody debris, and occasionally within clumps of moist ferns growing in waterfall spray zones. Fissures and crevices within the adjacent granite or limestone outcrops also harbor Inyo Mountains Salamanders and presumably serve as refugia during periods of unfavorable surface conditions.

Specific threats to Inyo Mountains Salamanders include mining activities, damage from livestock and feral burros, and water diversions.

Inyo Mountains Salamanders are a Species of Special Concern according to the California Department of Fish and Game and are listed as a Sensitive Species by the U.S. Forest Service and the Bureau of Land Management.

ORIGINAL ACCOUNT by Robert W. Hansen and David B. Wake; photograph by Gary Nafis.

Batrachoseps diabolicus Jockusch, Wake, and Yanev, 1998

Hell Hollow Slender Salamander

Hell Hollow Slender Salamanders are found on the western slope foothills of the Sierra Nevada of California, from the lower Merced River Canyon north to the American River, at elevations below 620 m.

Hell Hollow Slender Salamanders inhabit mostly mixed pine–oak woodland and chaparral communities, in areas where there may be extreme summer heat and drought. They are typically found beneath rocks, bark rubble, or downed logs, usually in areas that receive little direct sunlight during the winter, such as a narrow, shaded ravine. However, details about the distribution, habitat, and other aspects of the natural history of Hell Hollow Slender Salamanders are not well known.

Based upon limited sampling, populations from the Merced River Canyon appear stable. There is little information concerning the status of more northern populations.

Hell Hollow Slender Salamanders are not listed in the U.S. under any state or federal laws or regulations.

ORIGINAL ACCOUNT by Robert W. Hansen and David B. Wake; photograph by Gary Nafis.

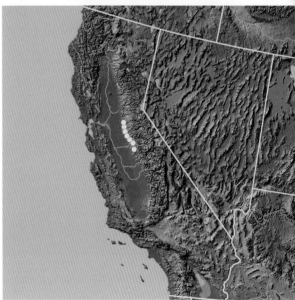

Hell Hollow Slender Salamander

Batrachoseps gabrieli Wake, 1996

San Gabriel Mountains Slender Salamander

San Gabriel Mountains
Slender Salamander

San Gabriel Mountains Slender Salamanders are known from scattered, discrete localities along the southern flanks of the San Gabriel and San Bernardino mountains of southern California, at elevations from 850 to 2380 m.

San Gabriel Mountains Slender Salamanders are talus specialists though, in general, their ecology remains poorly known. They have been discovered in and around stable talus accumulations under rotting logs, bark, downed branches, fern fronds, and rocks. These sites lay within mixed conifer forest consisting of various species of pines, white fir, big-cone spruce, incense cedar, and canyon live oak or at lower, drier, more exposed sites within chaparral communities associated with isolated stands of big-cone spruce.

San Gabriel Mountains Slender Salamanders are listed by the U.S. Forest Service as a Sensitive Species. All sites occur on public lands administered by Angeles and San Bernardino National Forests.

ORIGINAL ACCOUNT by Robert W. Hansen, Robert H. Goodman Jr., and David B. Wake; photograph by Gary Nafis.

Batrachoseps gavilanensis Jockusch, Yanev, and Wake, 2001

Gabilan Mountains Slender Salamander

Gabilan Mountains Slender Salamanders are known from many locations at elevations from near sea level to about 880 m in central coastal California, from the Santa Cruz Mountains around Monterey Bay southeast through the Gabilan Range and Sierra de Salinas to the Cholame Hills.

Gabilan Mountains Slender Salamanders are found in a variety of habitats from deeply shaded, moist redwood and mixed coniferous forests through oak woodland and chaparral to open grassland with widely scattered small oaks. The northwestern-most localities have cool, equable climates and rainfall that can exceed 100 cm annually, whereas the southeastern-most localities are hot and dry during the summer, with annual rainfalls of less than 20 cm.

Gabilan Mountains Slender Salamanders are moderately widespread throughout their range and appear to be common at a number of localities. Portions of their range occur on publicly owned lands or other large land holdings that are likely to remain relatively undisturbed for the foreseeable future. Aside from local extirpations associated with human development, there are no known critical conservation concerns.

Gabilan Mountains Slender Salamanders are not listed in the U.S. under any state or federal regulations.

ORIGINAL ACCOUNT by Robert W. Hansen and David B. Wake; photograph by Gary Nafis.

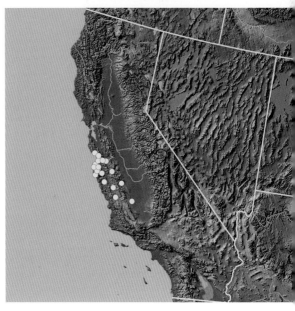

Gabilan Mountains
Slender Salamander

Batrachoseps gregarius Jockusch, Wake and Yanev, 1998

Gregarious Slender Salamander

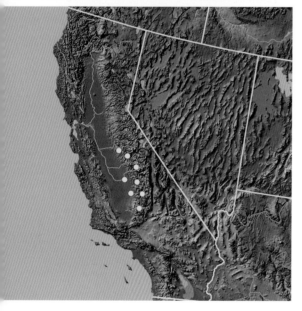

Gregarious Slender Salamander

Gregarious Slender Salamanders are found on the western slopes of the central and southern Sierra Nevada of California, from the southern boundary of Yosemite National Park south to just north of the Kern River, at elevations ranging from less than 100 m to about 1800 m.

Gregarious Slender Salamanders occur principally in the oak woodlands of the Sierra Nevada foothills, dominated by blue oak, interior live oak, and foothill pine, but also range into mixed forests of ponderosa pine, incense cedar, white fir, and black oak. At their northern range limits, Gregarious Slender Salamanders are abundant within a closed canopy conifer forest with sugar pine and giant sequoias. The southernmost population occupies arid rolling grassland with scattered rocks, under which Gregarious Slender Salamanders have been found. At some higher-elevation sites, Gregarious Slender Salamanders occur under or within woody debris adjacent to seeps or other moist places within coniferous forests. Gregarious Slender Salamanders may be especially abundant along riparian corridors and may even follow them onto the floor of the Central Valley.

There is little evidence of changes in distribution as compared with the historical range, and long-term trends in abundance are not evident. Some range-margin populations along the eastern edge of the Central Valley may have been affected by housing, agriculture, or other development.

Gregarious Slender Salamanders are not listed either federally or at the state level in the U.S.

ORIGINAL ACCOUNT by Robert W. Hansen and David B. Wake; photograph by Gary Nafis.

Batrachoseps incognitus Jockusch, Yanev, and Wake, 2001

San Simeon Slender Salamander

San Simeon Slender Salamanders are found in the Santa Lucia Range of central coastal California. Although generally found in the mountains, at elevations as high as 1000 m, they occur near sea level in the northwestern-most part of their range, where the mountains abruptly meet the ocean.

San Simeon Slender Salamanders occur in leaf litter in habitats ranging from open, predominantly yellow pine forest to closed canopy forests of laurel and sycamore. Near Rocky Butte in the San Simeon Creek drainage, San Simeon Slender Salamanders inhabit the forest edge but are absent from well-developed oak forest.

San Simeon Slender Salamanders do not have listed status at either the state or the federal level.

ORIGINAL ACCOUNT by Robert W. Hansen and David B. Wake; photograph by Gary Nafis.

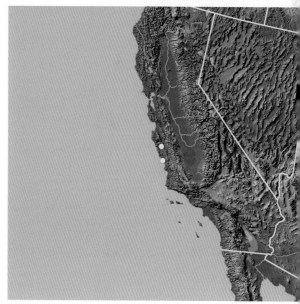

San Simeon Slender Salamander

Batrachoseps kawia Jockusch, Wake and Yanev, 1998

Sequoia Slender Salamander

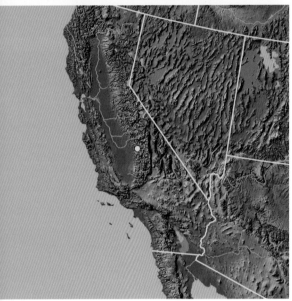

Sequoia Slender Salamander

The known distribution of Sequoia Slender Salamanders lies entirely within the Kaweah River drainage on the western slope of the Sierra Nevada of California, at elevations from 430 to 2200 m.

At lower elevations, Sequoia Slender Salamanders are present among moist, moss-covered talus on or at the base of a north-facing slope bordering a stream within foothill oak woodland. At somewhat higher elevations, they occur under fallen tree limbs and surface litter in mixed conifer forest. At the highest-elevation sites, Sequoia Slender Salamanders may be found under rocks on wet gravel or beneath wet logs resting on wet soil in shaded forest near brooks or perennial springs. Vegetation may include willows, currant, white fir, manzanita, sugar pine, incense cedar, giant sequoia, alders, and ferns. All known localities for Sequoia Slender Salamanders appear to have undergone little change in recent decades.

The majority of Sequoia Slender Salamander populations occur on public lands administered by the U.S. Forest Service or National Park Service. They thus enjoy some measure of habitat protection, although Forest Service lands are subject to various uses, including timber harvest and grazing.

Sequoia Slender Salamanders are not legally listed in California, nor are they federally listed in the U.S.

ORIGINAL ACCOUNT by Robert W. Hansen and David B. Wake; photograph by Gary Nafis.

Batrachoseps luciae Jockusch, Yanev, and Wake, 2001

Santa Lucia Mountains Slender Salamander

Santa Lucia Mountains Slender Salamanders are distributed throughout the northern Santa Lucia Mountains of coastal California from the Monterey Peninsula to as far south as 36°N latitude.

San Lucia Mountains Slender Salamanders are found predominantly in humid redwood and mixed coniferous forests. Inland, they occur mainly on wooded, north-facing slopes covered predominantly with tan bark oaks and maples. During favorable climatic conditions, San Lucia Mountains Slender Salamanders also can be found under suitable cover in open, disturbed habitats. In the Big Sur area, they have been found in wet, creeksides.

Santa Lucia Mountains Slender Salamanders are moderately widespread within their range and are abundant at a few sites. Aside from some local extirpations associated with human development, there are no known significant conservation concerns.

Santa Lucia Mountains Slender Salamanders are not listed under any state or U.S. federal laws or regulations.

ORIGINAL ACCOUNT by Robert W. Hansen and David B. Wake; photograph by Gary Nafis.

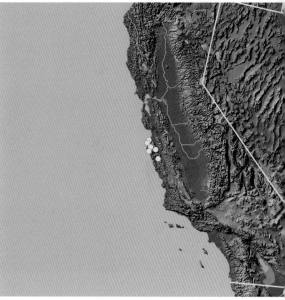

Santa Lucia Mountains
Slender Salamander

Batrachoseps major Camp, 1915

Southern California Slender Salamander

Southern California Slender Salamanders are found in southern California from the southern foothills of the Santa Monica, San Gabriel, and San Bernardino mountains south along the Pacific Coast and into Baja California (not shown). To the east, their range extends to the margins of the Colorado Desert and the city of Palm Springs. They also are found on Santa Catalina, North, Middle, and South Coronados, and Todos Santos Islands off the California Coast. Southern California Slender Salamanders have been introduced at Hanford, in the San Joaquin Valley of central California. They mainly occur at elevations from sea level to 700 m, but reach 1500 m on Mt. Palomar.

Prior to extensive modification of southern California native landscapes, Southern California Slender Salamanders occupied areas of coastal sage scrub, chaparral, and coast live-oak woodlands where summers are mild and dry and winters are wet, often with coastal fog. Southern California Slender Salamanders now tend to be associated with suburban landscapes that receive regular irrigation. On Santa Catalina Island, Southern California Slender Salamanders are found in coastal sage scrub.

Southern California Slender Salamander

Southern California Slender Salamanders appear to be tolerant of both heat and drought. They use gopher burrows, soil crevices, and earthworm tunnels for retreats. During periods of surface activity, Salamanders may be found under rocks, logs, boards, and other surface cover, as well as within moist leaf litter, mostly on coarse, well-drained substrates. Individuals occasionally are found climbing in low vegetation.

While formerly widespread and common, Southern California Slender Salamanders have been extirpated from much of their historical range because of habitat destruction.

Nevertheless, Southern California Slender Salamanders are not listed at either the state or federal level in the U.S.

ORIGINAL ACCOUNT by Robert W. Hansen and David B. Wake; photograph by Gary Nafis.

Batrachoseps minor Jockusch, Yanev and Wake, 2001

Lesser Slender Salamander

Lesser Slender Salamanders are restricted to the southern Santa Lucia Range northeast of Estero Bay in central coastal California, at elevations from 400 to 640 m.

Lesser Slender Salamanders appear to be restricted to areas that are either higher in elevation or wetter than surrounding areas, such as damp canyons surrounded by relatively dryer habitats. In such habitats, characterized by oak, sycamore, and laurel, Lesser Slender Salamanders occupy shaded slopes with deep leaf litter.

Lesser Slender Salamanders were once common but today are difficult to find. Although some areas within the historical range of this species have been modified for agriculture, particularly conversion to vineyards, ample habitat appears to remain, and there is no obvious reason for this decline in abundance.

Lesser Slender Salamanders are not listed in the U.S. under any state or federal laws or regulations.

ORIGINAL ACCOUNT by Robert W. Hansen and David B. Wake; photograph by Gary Nafis.

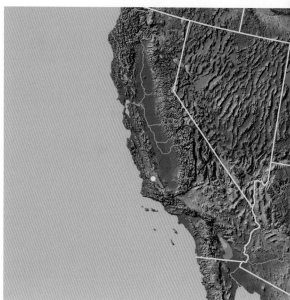

Lesser Slender Salamander

Batrachoseps nigriventris Cope, 1869

Black-bellied Slender Salamander

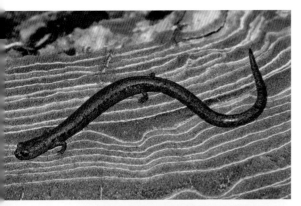

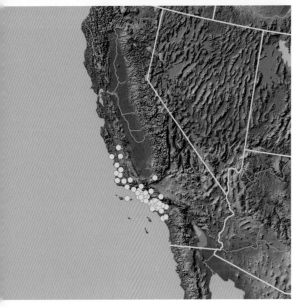

Black-bellied Slender Salamander

Black-bellied Slender Salamanders are found in the coastal mountains and valleys west of the Central Valley of California from the Santa Lucia Range south to the Santa Monica and San Gabriel mountains. They also are found in the Tehachapi Mountains at the southern and eastern margins of the Central Valley, in isolated upland areas surrounding the Los Angeles basin, and on Santa Cruz Island off the California Coast. They occur at elevations from near sea level to about 2260 m on Mt. Pinos.

Black-bellied Slender Salamanders occupy a wide range of habitats, from semiarid blue oak savannas to moist, oak-filled canyons and pine–fir forest. On the northern slopes of the Tehachapi Mountains, Black-bellied Slender Salamanders occur on semiarid, grassy slopes, whereas on Mt. Pinos, they are locally abundant in Jeffrey pine–white fir forests, especially in association with small streams and seepages. South of the Tehachapi Mountains, they mostly are found in coastal live-oak woodlands and/or chaparral. On Santa Cruz Island, Black-bellied Slender Salamanders occur under rocks and scattered debris in open grassland; under rocks, fallen branches, and in leaf litter in oak woodland; in and under rotting branches and logs as well as under rocks; in surface litter in pine forests; and under superficial surface cover in chaparral.

Long-term trends in the abundance of Black-bellied Slender Salamanders are not apparent.

Black-bellied Slender Salamanders are not listed in the U.S. under any state or federal regulations.

ORIGINAL ACCOUNT by Robert W. Hansen and David B. Wake; photograph by Brad Moon.

Batrachoseps pacificus (Cope, 1865)

Channel Islands Slender Salamander

Channel Islands Slender Salamanders are restricted to the northern Channel Islands of East Anacapa, Middle Anacapa, West Anacapa, Santa Cruz, Santa Rosa, and San Miguel off the Pacific Coast of south central California. They range in elevation from sea level to around 430 m on Santa Cruz Island.

Channel Islands Slender Salamanders occur under rocks and logs, especially near streams and in grassland, coastal sage scrub, chaparral, riparian oak woodlands, and pine forest. Dense populations have been found in open areas near the ocean, and Channel Islands Slender Salamanders may be abundant under driftwood on sand within 50 to 60 m of the sea. Periods of surface activity correspond generally to the rainy season, especially in drier inland valleys of Santa Cruz Island. However, the moderating influence of cool, marine air combined with daily fog extends activity throughout the summer.

There are no known threats to Channel Islands Slender Salamanders. They appear to occupy all parts of their potential range, and there is no indication of any changes in abundance. Santa Rosa and Santa Cruz Islands have a variety of introduced mammals present, but the other four islands where Channel Islands Slender Salamanders live are managed entirely as natural areas and appear to provide good habitat. Channel Islands Slender Salamanders are not listed at either the state or federal level in the U.S.

Channel Islands Slender Salamander

ORIGINAL ACCOUNT by Robert W. Hansen, David B. Wake, and Gary M. Fellers; photograph by Gary Nafis.

Batrachoseps regius Jockusch, Wake, and Yanev, 1998

Kings River Slender Salamander

Kings River Slender Salamander

Kings River Slender Salamanders are known from three sites in and near the Kings River drainage on the western slope of the Sierra Nevada of California at elevations of 335 to 440 m, 610 m, and 2470 m. All of the known populations of Kings River Slender Salamanders appear to be quite localized and apparently have been isolated from one another for a long time.

At lower elevations, Kings River Slender Salamanders live within a mixed pine–oak/chaparral habitat characterized by interior live oak, blue oak, foothill pine, and western redbud. The lone high-elevation site is located within a moist coniferous forest of lodgepole pine and red fir. Kings River Slender Salamanders have been found under scattered granitic rocks or downed rotted logs or within leaf litter at the base of shaded, damp slopes and ravines.

All localities for Kings River Slender Salamanders occur on public lands administered by the U.S. Forest Service or National Park Service. The lower Kings River sites are located immediately adjacent to a road and likely would be affected by road construction.

Kings River Slender Salamanders are not listed in California, nor are they federally listed in the U.S.

ORIGINAL ACCOUNT by Robert W. Hansen and David B. Wake; photograph by Gary Nafis.

Batrachoseps relictus Brame and Murray, 1968

Relictual Slender Salamander

Relictual Slender Salamanders are restricted to the west slopes of the southern Sierra Nevada of California, ranging from the lower Kern River Canyon to the highlands drained by the Tule and Kern rivers. In the lower Kern River Canyon, Relictual Slender Salamanders occur at elevations from 485 to 730 m. In other areas, they occur at elevations of 1125 to 2440 m.

Relictual Slender Salamanders are associated with downed logs and bark rubble in moist conifer forest containing Ponderosa pine, sugar pine, incense cedar, white fir, and black oak. They frequently are found near seepages and springs where surface moisture persists through the summer. At high elevations, Relictual Slender Salamanders are found in a large seepage surrounded by a dense forest of Jeffrey pine and white fir.

The construction of State Route 178 severely impacted seepages and springs in the lower Kern River Canyon, which once harbored Relictual Slender Salamanders; this loss of habitat may have initiated local population declines. Aside from the lower Kern River Canyon, though, remaining populations of Relictual Slender Salamanders appear to be stable.

Relictual Slender Salamanders are a Species of Special Concern according to the California Department of Fish and Game, as a Sensitive Species by the U.S. Forest Service , and as a Species of Concern by the U.S. government.

Relictual Slender Salamander

ORIGINAL ACCOUNT by Robert W. Hansen and David B. Wake; photograph by Gary Nafis.

Batrachoseps robustus Wake, Yanev, and Hansen, 2002

Kern Plateau Salamander

Kern Plateau Salamander

Kern Plateau Salamanders are restricted to the southeastern Sierra Nevada of California. Their range consists of the Kern Plateau at elevations from 1700 to 2800 m, the eastern slopes of the Sierra Nevada draining into Owens and Indian Wells valleys from 1430 to 2440 m, and the Scodie Mountains from 1980 to 2025 m.

Kern Plateau Salamanders occur in ecologic settings ranging from high-elevation coniferous forests to semiarid pinyon pine and sagebrush country, provided there is sufficient moisture. During favorable periods of surface moisture and temperature, Kern Plateau Salamanders may be found under rocks, under or within downed logs, or among bark rubble. Hatchling Kern Plateau Salamanders often are found under smaller cover objects that are not used by adults.

Kern Plateau Salamanders have been found to be surprisingly widespread on the Kern Plateau, especially on the more humid portions, as well as being present in virtually every stream-bearing canyon on the eastern flank of the Sierra Nevada within their range.

Nearly all populations of Kern Plateau Salamanders occur on public lands administered by the U.S. Forest Service or the U.S. Bureau of Land Management. Some sites have been affected by road construction, timber-harvesting activities, or forest-fire-suppression efforts.

Kern Plateau Salamanders are listed by the U.S. Forest Service as a Sensitive Species.

ORIGINAL ACCOUNT by Robert W. Hansen and David B. Wake; photograph by Gary Nafis.

Batrachoseps simatus Brame and Murray, 1968

Kern Canyon Slender Salamander

Kern Canyon Slender Salamanders are known only from the Kern River drainage in the southern Sierra Nevada of California from elevations of 1920 m down to 330 m in the lower Kern River Canyon.

Kern Canyon Slender Salamanders occur mostly in tributary canyons or ravines at the base of a north-facing slope within pine–oak woodland. These areas receive little or no direct sunlight during the winter and are dominated by foothill pine, live oak, sycamore, California buckeye, Fremont cottonwood, and willow. At higher elevations, Kern Canyon Slender Salamanders are associated with talus. Kern Canyon Slender Salamanders have been found under rocks, under and within logs, and in moist oak and sycamore litter.

Nearly all the known populations of Kern Canyon Slender Salamanders occur on public lands administered by the Sequoia National Forest. Cattle grazing has severely degraded the habitat of Kern Canyon Slender Salamanders, particularly in narrow ravines.

Kern Canyon Slender Salamanders are listed as Threatened by the State of California, as a Species of Concern by the U.S. government, and as a Sensitive Species by the U.S. Forest Service.

ORIGINAL ACCOUNT by Robert W. Hansen and David B. Wake; photograph by Gary Nafis.

Kern Canyon Slender Salamander

Batrachoseps stebbinsi Brame and Murray, 1968

Tehachapi Slender Salamander

Tehachapi Slender Salamander

Tehachapi Slender Salamanders are known from two small areas in the Tehachapi Mountains of south central California: Caliente Canyon at elevations between 550 and 790 m and several canyons from Tejon Canyon southwest to Fort Tejon at elevations between 945 and 1430 m.

Tehachapi Slender Salamanders are confined to seasonally shaded, north-facing slopes of canyons located in otherwise arid to semiarid terrain. Within these canyons, Tehachapi Slender Salamanders are associated with talus, scattered rocks, or downed wood amid foothill pine, interior live oak, canyon oak, blue oak, Fremont cottonwood, sycamore, and California buckeye. Juvenile Tehachapi Slender Salamanders rarely are found, suggesting that hatching occurs in spring as surface activity declines and juveniles may remain well underground.

Some sites occupied by Tehachapi Slender Salamanders have been affected by road construction, mining, and cattle grazing and potentially by flood-control projects. Portions of the Tehachapi Mountains are experiencing rapid human population growth, with much development occurring in the foothills.

Tehachapi Slender Salamanders are listed as Threatened by the State of California, as a Sensitive Species by the U.S. Forest Service, and as a Species of Concern by the U.S. government.

ORIGINAL ACCOUNT by Robert W. Hansen and David B. Wake; photograph by Gary Nafis.

Batrachoseps wrighti (Bishop, 1937)

Oregon Slender Salamander

Oregon Slender Salamanders occur in western Oregon from the south side of the Columbia River Gorge southward in the Cascade Mountains to the vicinity of Odell Lake. Populations typically occur on the west slopes of the Cascades at elevations from about 15 m to 1340 m.

Oregon Slender Salamanders are most often found within large, well-decayed logs and stumps in old growth and mature forests. They occasionally are found under large rocks that are moss-covered and in stabilized talus. Oregon Slender Salamanders are active after snowmelt in the spring, but at other times of the year inhabit underground burrows or sites deep within large logs.

Declines in the abundance of Oregon Slender Salamanders have been associated with the clear-cutting of ancient and mature forests.

Oregon Slender Salamanders are considered by the State of Oregon to be a Sensitive Species of vulnerable status and are listed as Protected, which means that a permit is required to possess or collect them.

ORIGINAL ACCOUNT by R. Bruce Bury; photograph by Gary Nafis.

Oregon Slender Salamander

Desmognathus abditus Anderson and Tilley, 2003

Cumberland Dusky Salamander

Cumberland Dusky Salamanders occur on the Cumberland Plateau of Tennessee from just south of the Cumberland Mountains north to Walden Mountain.

Cumberland Dusky Salamanders are found under rocks beside small streams or under moss and debris on vertical rock surfaces behind cascades, usually within 1 m of the water.

The abundance of Cumberland Dusky Salamanders and the viability of their populations are unknown.

Cumberland Dusky Salamanders do not receive protection by the state of Tennessee or by the U.S. federal government.

ORIGINAL ACCOUNT by Michael J. Lannoo; photograph by Nathan Haislip.

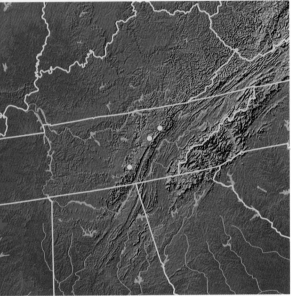

Cumberland Dusky Salamander

Desmognathus aeneus Brown and Bishop, 1947

Seepage Salamander

Seepage Salamanders are distributed in the Blue Ridge and adjacent Piedmont region of southwestern North Carolina, northwestern South Carolina, southeastern Tennessee, northern Georgia, and north central Alabama as well as in the Fall Line Hills region of Alabama. Another, apparently disjunct, population is present in the Piedmont of northeastern Georgia.

Seepage Salamanders are terrestrial and live at the interface between the leaf or leaf-mold layer and the underlying soil in the vicinity of seepages and small streams in heavily shaded hardwood or mixed forests. When active, they are nocturnal, typically remain under surface cover, and generally neither climb nor actively burrow. It is therefore odd that Seepage Salamanders have been observed climbing on grasses and bushes and jumping from branch to branch at a locality in central Alabama.

Logging activities have been responsible for the extirpation of some Alabama populations of Seepage Salamanders.

In Alabama, Seepage Salamanders are currently ranked as S2, Imperiled, because of their rarity, and their present status in North Carolina is SR, or Significantly Rare. Georgia accords Seepage Salamanders no legal status, but in Tennessee, they are regarded as a species in need of management. The U.S. Fish and Wildlife Service previously listed Seepage Salamanders as a Category 2 candidate for listing.

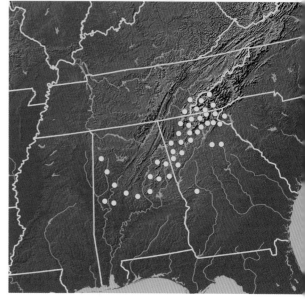

Seepage Salamander

ORIGINAL ACCOUNT by Julian R. Harrison; photograph by Dante Fenolio.

Desmognathus apalachicolae Means and Karlin, 1994

Apalachicola Dusky Salamander

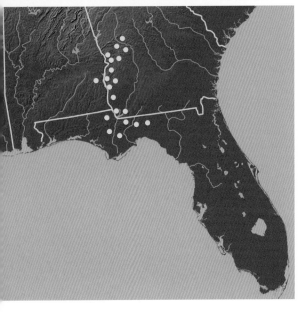

Apalachicola Dusky Salamander

Apalachicola Dusky Salamanders are confined principally to tributaries of the Apalachicola and Chattahoochee rivers of Florida, Alabama, and Georgia, though populations are also known from the Chipola River basin, the Ochlockonee River basin, and the upper Choctawhatchee River drainage.

Apalachicola Dusky Salamanders inhabit first-order streams in gully-eroded ravine valley heads or steepheads. During the day, Apalachicola Dusky Salamanders are under logs, rocks, debris, or leaf packs in moist ravines and usually reside with their bodies immersed in water with only their heads protruding. When suddenly exposed, Apalachicola Dusky Salamanders usually dive into water to escape. At night, adults forage in leaf litter or over the moist soil at the edge of seepages and on the banks of small streams. They are rarely found more than 2 m from water.

Apalachicola Dusky Salamander populations seem to be stable, in part because of their proclivity for deep, shaded, wet ravines, which are unsuitable for human development.

Apalachicola Dusky Salamanders are potentially protected in Torreya State Park and the Apalachicola Bluffs and Ravines Preserve of The Nature Conservancy in Florida, and in Kolomoki Mounds State Park in Georgia. However, they are not legally listed by any jurisdiction.

ORIGINAL ACCOUNT by D. Bruce Means; photograph by Aubrey Huepel.

Desmognathus auriculatus (Holbrook, 1838)

Southern Dusky Salamander

Southern Dusky Salamanders inhabit the Coastal Plain of the southeastern U.S. from extreme southeastern Virginia to midpeninsular Florida, then west to the Trinity River basin of eastern Texas.

Southern Dusky Salamanders are found in such locations as shallow water under logs along creek edges, in swampy mucklands or ravines around seepages, in muddy bottomland swamps and sloughs, among debris at the edge of mucky floodplain sloughs, at mucky edges of swampy lakes, or along the edges of springs and spring-runs under leaf mold, logs, or mats of marginal aquatic vegetation. Female Southern Dusky Salamanders attend their unhatched eggs and hatchlings under logs or in small crevices in wet mud in such places as cypress swamps.

There has been an alarming decline in the number of localities where Southern Dusky Salamanders may now be found. They have severely declined in the Florida Parishes of Louisiana and they have been extirpated from many sites in Florida, Georgia, South Carolina, North Carolina, and Virginia.

Southern Dusky Salamanders are not listed in any of the states where they occur, nor are they federally listed in the U.S.

ORIGINAL ACCOUNT by D. Bruce Means; photograph by Brad Moon.

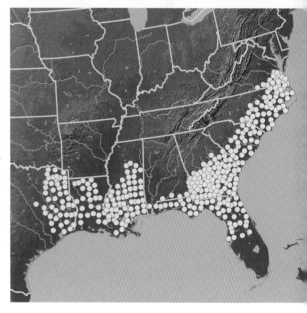

Southern Dusky Salamander

Desmognathus brimleyorum Stejneger, 1895

Ouachita Dusky Salamander

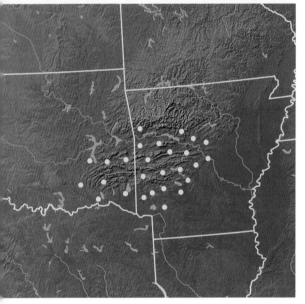

Ouachita Dusky Salamander

Ouachita Mountain Dusky Salamanders occur in highland areas in western Arkansas and eastern Oklahoma south of the Arkansas River, including the Ouachita Mountains and the Potato Hills. Ouachita Mountain Dusky Salamanders are also found in small streams in the rugged topography south of Kiamichi Mountain. One population is known north of the Arkansas River in Arkansas.

Ouachita Mountain Dusky Salamanders are found most often and abundantly in first- and second-order mountain streams and adjacent ravine woodlands, especially where there is a rocky, gravelly, porous substrate. Adult Ouachita Mountain Dusky Salamanders may venture up to several tens of meters away from the streamside to live among coarse, rocky talus. At night, they sit with their heads and anterior bodies protruding from crevices between rocks.

Ouachita Mountain Dusky Salamanders may be sensitive to silvicultural practices, such as clear-cutting, that may increase sedimentation in streams, alter the hydrology of mountain brook watersheds, or deliver pesticides, herbicides, or fertilizers to aquatic habitats. The fish-bait industry may also affect Ouachita Mountain Dusky Salamanders because of the harvest and transport of animals between populations. Most of their geographic range, however, is on publicly owned lands, including the Ouachita National Forest, the Ozark National Forest, Hot Springs National Park, and several state parks in both Arkansas and Oklahoma.

Ouachita Dusky Salamanders are not listed either federally or at the state level in the U.S.

ORIGINAL ACCOUNT by D. Bruce Means; photograph by John Clare.

Desmognathus carolinensis Dunn, 1916

Carolina Mountain Dusky Salamander

Carolina Mountain Dusky Salamanders are found on the Blue Ridge, Black, Bald, and Unaka Mountains of eastern Tennessee and western North Carolina from Linville Falls and McKinney Gap to the Pigeon River Valley. Populations range to the peaks of the highest mountains in the area at elevations of approximately 2000 m.

Carolina Mountain Dusky Salamanders are abundant in seepage areas, on wet rock faces, and in forest-floor litter in association with streams, particularly headwater seepages. Individuals living at moist, high-elevation sites may roam on the forest floor far from streams, whereas individuals at low elevations stay in close association with seeps and streams. Carolina Mountain Dusky Salamanders are largely nocturnal, remaining beneath cover during the day and emerging to feed at night. They may be active during the day when it is heavily overcast. On rainy or foggy nights, Carolina Mountain Dusky Salamanders frequently climb understory plants and tree trunks and perch over a meter above the forest floor.

Because of their reliance on moist habitats, the greatest potential threat to the survival of Carolina Mountain Dusky Salamanders is likely to be habitat desiccation resulting from the removal of protective forest canopy through timber harvesting.

Carolina Mountain Dusky Salamanders are not listed under any state or federal laws or regulations in the U.S.

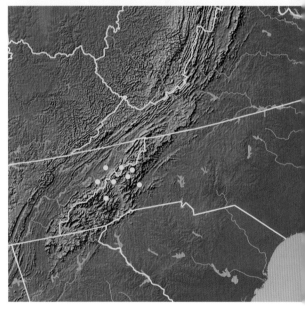

Carolina Mountain Dusky Salamander

ORIGINAL ACCOUNT by Carlos D. Camp and Stephen G. Tilley; photograph by Gary Nafis.

Desmognathus conanti Rossman, 1958

Spotted Dusky Salamander

The geographic distribution of Spotted Dusky Salamanders extends from the confluence of the Cumberland and Ohio rivers southeast into northwestern South Carolina, as far south as the Gulf Coast from Apalachee Bay to the Mississippi River, and west to western Louisiana and southern Arkansas.

Spotted Dusky Salamanders may be found in a variety of damp locations, including seepage areas, cold springs, and the edges of small rocky streams in deeply shaded, heavily wooded ravines. There, the salamanders reside under rocks, damp leaf litter, or rotting logs.

Spotted Dusky Salamanders are adversely affected by stream disturbance associated with urbanization. Siltation and sedimentation of their small stream habitats from runoff following construction and farming probably have extirpated or severely reduced populations of Spotted Dusky Salamanders throughout their range.

Spotted Dusky Salamanders lack listed status at either state or federal levels in the U.S.

ORIGINAL ACCOUNT by D. Bruce Means and Ronald M. Bonett; photograph by Mike Redmer.

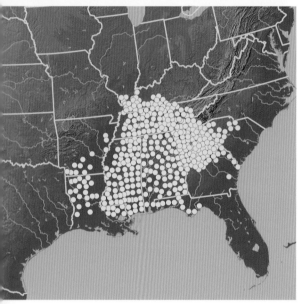

Spotted Dusky Salamander

Desmognathus folkertsi Camp, Tilley, Austin, and Marshall, 2002

Dwarf Black-bellied Salamander

Dwarf Black-bellied Salamanders are currently known from tributaries of Wolf Creek and Helton Creek in the Nottely River drainage of the Blue Ridge in Georgia and adjacent North Carolina.

Dwarf Black-bellied Salamanders are to be found in high-gradient, turbulent first- and second-order streams. They are most abundant in shallow areas of streams, where they take cover under rocks. At night they occasionally will wander away from cover but usually remain with only their heads and anterior parts of their bodies protruding from cover sites. Juvenile Dwarf Black-bellied Salamanders may be found among wet leaves at the margins of streams. Dwarf Black-bellied Salamanders are abundant in the two streams where they occur.

Dwarf Black-bellied Salamanders are not legally listed at present.

ORIGINAL ACCOUNT by Carlos D. Camp and Stephen G. Tilley; photograph by Gary Nafis.

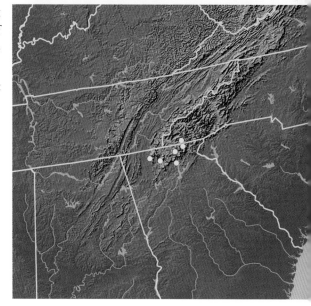

Dwarf Black-bellied Salamander

Desmognathus fuscus (Rafinesque, 1820)

Northern Dusky Salamander

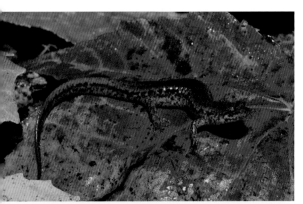

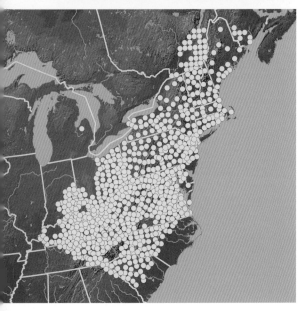

Northern Dusky Salamander

Northern Dusky Salamanders occur in the Niagara Gorge in Ontario, in southeastern Québec, in southern New Brunswick south to western South Carolina and west to western Kentucky. The southern geographic limit of Northern Dusky Salamanders is a poorly delineated contact zone between them and Spotted Dusky Salamanders that extends southeast from the Cumberland River in western Kentucky to western South Carolina. There is a disjunct record of Northern Dusky Salamanders in the Saginaw Peninsula of Michigan.

Northern Dusky Salamanders are found along the margins of small woodland brooks, on seepage hillsides, in shallow weed-choked streams with sandy or gravelly bottoms, in low boggy places, along leaf-littered trickles, in spring banks with constantly moist soil, or in the beds of ephemeral streams. They tend to be hidden under stones, flat rocks, logs, bark, and other debris either on the ground or partially submerged in streams. They may often be found in the mouths of caves, but not within caves.

Urbanization has extirpated many populations of Northern Dusky Salamanders, but they are wide-ranging and ubiquitous in small stream valleys. The absence of larvae of Northern Dusky Salamanders from streams draining coal strip mines appears to be caused by siltation and high metal concentrations.

New Brunswick lists North Dusky Salamanders as Sensitive and Québec considers them to be a species likely to be designated threatened or vulnerable. North Dusky Salamanders in Ontario are listed both federally and provincially as Endangered. In the U.S., North Dusky Salamanders are not listed under any state or federal regulations.

ORIGINAL ACCOUNT by D. Bruce Means; photograph by John Clare.

Desmognathus imitator (Dunn, 1927)

Imitator Salamander

Imitator Salamanders are found in a restricted area of the Great Smoky, Plott Balsam, and Balsam mountains of eastern Tennessee and western North Carolina at elevations above 900 m. A population on Waterrock Knob occurs on wet rock faces above 1650 m.

Imitator Salamanders are found along streamsides and on wet rock faces associated with the forest floor at high elevations in spruce fir and northern hardwood forests. Juveniles can be abundant in headwater seeps.

Most of the geographic range of Imitator Salamanders falls within the boundaries of the Great Smoky Mountains National Park; this largely protects them from such threats as mining and the harvesting of timber. The greatest long-term threat to Imitator Salamanders may result from acid precipitation that may fall on high elevation sites in the Great Smoky Mountains and other areas of the southern Appalachians.

Imitator Salamanders are not listed in the U.S. under any state or federal laws or regulations.

ORIGINAL ACCOUNT by Carlos D. Camp and Stephen G. Tilley; photograph by John Clare.

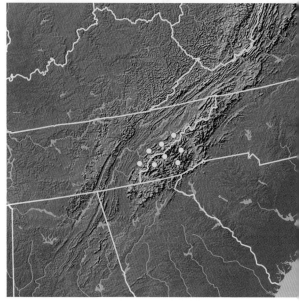

Imitator Salamander

Desmognathus marmoratus (Moore, 1899)

Shovel-nosed Salamander

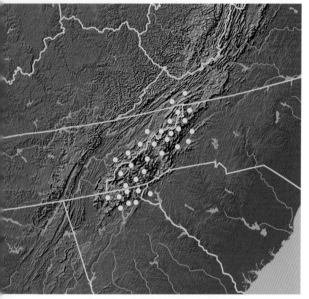

Shovel-nosed Salamander

Shovel-nosed Salamanders are found in the Blue Ridge and Great Smoky Mountains from southwestern Virginia southwest through eastern Tennessee, western North Carolina, extreme northwestern South Carolina, and into northern Georgia at elevations from 300 to 1680 m. Drainage patterns, and consequentially rate of stream flow, rather than elevation, apparently limit distribution. Populations are scattered, especially at the northeastern and southwestern extremes of the range. As Shovel-nosed Salamanders commonly were collected and sold as fish bait in Georgia during the mid-20th century, they have been introduced into areas outside their historical range.

Shovel-nosed Salamanders inhabit cool, well-oxygenated, second- and third-order streams or low-gradient first-order streams in areas with rocks, loose gravel, and moderate- to fast-flowing water, preferring rapids and riffles rather than pools. They hide under cover objects during the day and emerge to feed at night. Shovel-nosed Salamanders may move onto land during rainy weather and even onto branches just above flowing water.

Shovel-nosed Salamanders are vulnerable to degradation of the rocky, flowing streams they depend upon and so are largely absent from streams that have been heavily silted. They have been eliminated from many areas because of the impoundment of streams. Because they feed on aquatic insects, pollution or other degrading factors that affect insect populations may also affect populations of Shovel-nosed Salamanders. They are also vulnerable to acidification of streams and contamination of streams by heavy metals. Shovel-nosed Salamanders are no longer important to the fish-bait industry in Georgia.

Shovel-nosed Salamanders are not listed in the U.S. under any state or federal laws or regulations.

ORIGINAL ACCOUNT by Carlos D. Camp and Stephen G. Tilley; photograph by William Leonard.

Desmognathus monticola Dunn, 1916

Seal Salamander

Seal Salamanders range throughout the central and southern Appalachians from western Pennsylvania to central Alabama. Disjunct populations occur in the Coastal Plain of southern Alabama and the Florida Panhandle. Seal Salamanders are most commonly found below elevations of 1219 to 1372 m, although they occur as high as 1555 m. Because Seal Salamanders have widely been used as fish bait, they may have been introduced into new areas by anglers.

Seal Salamanders are found in hardwood forests in association with seepages and small- to medium-sized streams containing cool, well-aerated water. They live under rocks or logs, or in burrows, in association with streambanks and uninundated portions of streambeds, rather than in the stream channel proper. They also occasionally use crevices associated with wet cliffs as refugia. Juvenile Seal Salamanders are generally found closer to the water in streambeds, and under smaller objects, than adults. They occasionally are found wandering on the faces of wet cliffs. Following heavy rains, Seal Salamanders may forage in surrounding forest, occasionally climbing on tree trunks 1 to 2 m above the ground.

Seal Salamanders remain abundant over much of their geographic range. However, certain populations of Seal Salamanders may have experienced declines due to exploitation as "spring lizards" sold as fish bait. Stream acidification as a consequence of mine drainage may negatively affect some populations, as may timber-harvesting techniques that result in increased rates of evaporative water loss through the removal of protective forest canopy.

Seal Salamanders lack listed status at both state and federal levels.

ORIGINAL ACCOUNT by Carlos D. Camp and Stephen G. Tilley; photograph by Brad Moon.

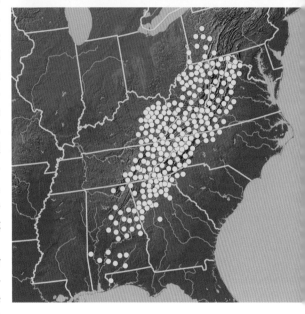

Seal Salamander

Desmognathus ochrophaeus Cope, 1859

Allegheny Mountain Dusky Salamander

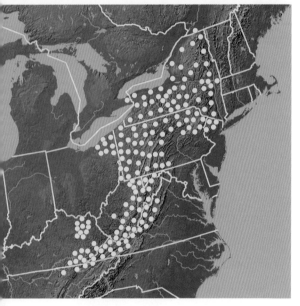

Allegheny Mountain
Dusky Salamander

The range of Allegheny Mountain Dusky Salamanders extends from southwestern Virginia west into the Cumberland Mountains and Cumberland Plateau of northeastern Tennessee and southeastern Kentucky, north through the Allegheny Plateau and Allegheny Mountains and through the Adirondack Mountains into extreme southern Québec. They also occur on the Canadian side of the Niagara River Gorge in Ontario. While Allegheny Mountain Dusky Salamanders will range in elevation from 975 to 1400 m, they appear to be most abundant between 975 and 1036 m.

Allegheny Mountain Dusky Salamanders occupy moist woodlands, seepage areas, wet rock faces, and small streams throughout most of their range. Adults are found under rocks, leaves, bark, and logs and can be found in forests some distance upslope from streams. They can be abundant in moist ravines, moist old logging roads, and close to seepages. Juveniles are observed more frequently under and between damp leaves on the forest floor and at the edges of first-order streams.

Allegheny Mountain Dusky Salamanders are common in the central and northern Appalachian Mountains. They occupy a variety of forested habitats and appear to be able to survive silvicultural impacts and forest fragmentation.

Allegheny Mountain Dusky Salamanders are listed as *"menacée"* (endangered) under Québec's *"Loi sur les espèces menacées ou vulnérables"* and as Endangered under Ontario's Endangered Species Act. They are listed as Threatened in Québec and Endangered in Ontario under the Canadian Species at Risk Act. They are not listed in the U.S.

ORIGINAL ACCOUNT by Thomas K. Pauley and Mark B. Watson; photograph by Michael Graziano.

Desmognathus ocoee Nicholls, 1949

Ocoee Salamander

Ocoee Salamanders occur in the southwestern Blue Ridge and adjacent Piedmont regions, including the Blue Ridge and Great Smoky Mountains and the Appalachian Plateau of northeastern Alabama, as well as the gorges of the Hiwassee, Ocoee, Tugaloo, and Tallulah Rivers. Ocoee Salamanders occur over a great range of elevations, from low-lying gorges to mountain tops.

Ocoee Salamanders have a strong affinity for the headwaters of first-order streams in montane regions and occur in seepages, on wet rock faces, and under cover along the edges and in the beds of streams. They will move away from the streambeds under humid conditions, and in the wetter forests of higher elevations they may be quite terrestrial. Ocoee Salamanders are substantially more abundant in mature cove hardwood forests than in younger stands. They are also more abundant in cove forests with a significant amount of emergent rock.

Ocoee Salamanders are among the most common salamanders of the southern Appalachians. Because of their reliance on moist habitats, the greatest potential threat to them is probably the desiccation of habitats caused by the removal of protective forest canopy through the harvesting of timber. Ocoee Salamanders occupying high-elevation sites may also be vulnerable to the effects of acid precipitation.

Ocoee Salamanders are not listed either federally or at the state level in the U.S.

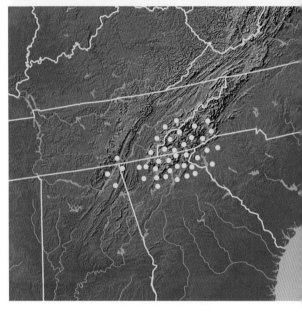

Ocoee Salamander

ORIGINAL ACCOUNT by Carlos D. Camp and Stephen G. Tilley; photograph by Dante Fenolio.

Desmognathus orestes Tilley and Mahoney, 1996

Blue Ridge Dusky Salamander

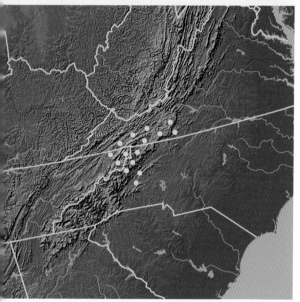

Blue Ridge Dusky Salamander

Blue Ridge Dusky Salamanders occur in the Blue Ridge from southwestern Virginia into northwestern North Carolina and northeastern Tennessee up to the tops of the highest mountains within this range.

Blue Ridge Dusky Salamanders inhabit the headwaters of first-order streams, where they are found associated with wet rock faces, seepages, and forest-floor habitats in the vicinity of streams and seeps. Adults are nocturnal and spend the day under cover objects, although they can be active during the day under overcast conditions. Juveniles are commonly associated with seepages of headwater streams, where they may be found on stream banks, in streambeds, and on wet rock faces.

Blue Ridge Dusky Salamanders are abundant at high elevations. The removal of protective forest canopy through timber harvesting, which results in desiccation of habitats, is a threat to Blue Ridge Dusky Salamanders because of their reliance on moist habitats. Populations at low elevations may take many years to recover from intensive timber harvesting. Populations of Blue Ridge Dusky Salamanders that occupy high-elevation sites in the southern Appalachians may also be vulnerable to the effects of acid precipitation.

Blue Ridge Dusky Salamanders are not listed under any state or federal laws or regulations in the U.S.

ORIGINAL ACCOUNT by Carlos D. Camp and Stephen G. Tilley; photograph by Gary Nafis.

Desmognathus organi Crespi, Browne, and Rissler, 2010[14]

Northern Pigmy Salamander

Northern Pigmy Salamanders occur north and east of the French Broad River in the mountains of southwestern Virginia, western North Carolina, and adjacent Tennessee. They are generally limited to elevations above 1100 m.

Northern Pigmy Salamanders are highly terrestrial and may be found beneath small logs or stones, or under the bark of stumps or trees. At higher elevations, Northern Pigmy Salamanders are characteristically resident in spruce–fir forests but at lower elevations will inhabit damp hardwood forests. On especially wet nights, Northern Pigmy Salamanders may be found on leaves some distance above the ground.

Logging activities and increased recreational development could potentially threaten Northern Pigmy Salamanders in portions of their range, particularly at lower elevations.

Northern Pigmy Salamanders are not listed in any of the states where they occur, nor are they federally listed in the U.S.

ACCOUNT by Julian R. Harrison; photograph by Erica Crespi.

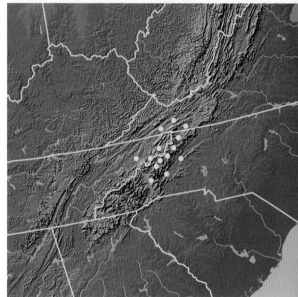

Northern Pigmy Salamander

Desmognathus planiceps Newman, 1955

Flat-headed Salamander

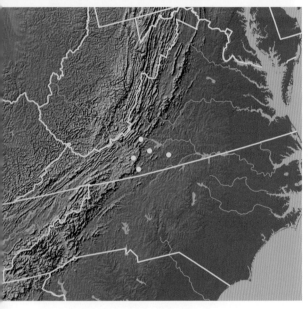

Flat-headed Salamanders occur in the upper reaches of the Roanoke River and Dan River drainages of southwest Virginia, mainly on the Atlantic slopes of the Blue Ridge Divide from just south of Roanoke to near the Pinnacles of Dan at elevations between 333 and 864 m. They also occur at 253 m near Smith Mountain in the Virginia Piedmont.

Flat-headed Salamanders occupy rocky, first- and second-order streams in deciduous woodlands. Individuals seek refuge under rocks, logs, rubbish, or other sheltering objects at the water's edge during daylight hours.

Flat-headed Salamanders are vulnerable to sources of stream degradation, including logging operations on steep mountain slopes.

Flat-headed Salamanders are not listed by either Virginia or U.S. statutes.

ORIGINAL ACCOUNT by Stephen G. Tilley; photograph by Richard D. Bartlett.

Flat-headed Salamander

Desmognathus quadramaculatus Holbrook, 1840

Black-bellied Salamander

Black-bellied Salamanders range through the Appalachian Mountains from northern Georgia, northwestern South Carolina, western North Carolina, eastern Tennessee, and southeastern Virginia to West Virginia. In the Great Smoky Mountains of Tennessee, they have been found at elevations from 375 to 1725 m. Black-bellied Salamanders have probably been introduced into new drainages by fisherman releasing individuals used as bait.

Black-bellied Salamanders inhabit cool mountain streams, especially small-volume, high-gradient, perennial streams. They are associated with clean water and large cobbles and flat rocks, and they may also use burrows in stream banks as refugia or for ambushing prey. Juvenile Black-bellied Salamanders tend to be farther into the middle of streams than adults, where they occupy smaller substrates and avoid refugia containing larger salamanders. Juveniles will flee from large conspecifics and spend more time actively foraging outside of refugia than do adults.

Stream pollution caused by acid mine drainage and sewage, clear-cut forestry practices, and overcollection for fish bait are potential threats to Black-bellied Salamanders, though there has apparently been no substantive changes in their abundance in undisturbed locations in the Appalachian Mountains.

Black-bellied Salamanders are listed as a Species of Special Concern in West Virginia.

ORIGINAL ACCOUNT by Mark B. Watson, Thomas K. Pauley, and Carlos D. Camp; photograph by Michael Graziano.

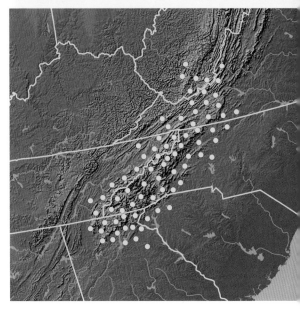

Black-bellied Salamander

Desmognathus santeetlah Tilley, 1981

Santeetlah Dusky Salamander

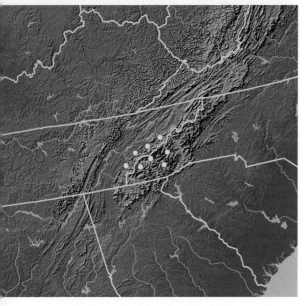

Santeetlah Dusky Salamander

Santeetlah Dusky Salamanders occur at elevations above 1000 m in the Unicoi, Great Smoky, Great Balsam, and Cheoah mountains in the southern Appalachians of western North Carolina and adjacent Tennessee.

Santeetlah Dusky Salamanders occur under cover along small streams and seepages in moist, high-elevation forests. They may go several meters into the surrounding forest but usually occur in or within a few centimeters of surface water. They also may be found on wet cliff faces.

A substantial portion of the geographic range of Santeetlah Dusky Salamanders falls within the boundaries of the Great Smoky Mountains National Park, which therefore protects them from such environmental disturbances as mining and timber harvesting. The stream-headwater habitats of Santeetlah Dusky Salamanders can be degraded by logging, road building, construction, and the activities of feral hogs, which is particularly important in the Unicoi Mountains. Populations of Santeetlah Salamanders occupying high-elevation sites in the southern Appalachians may also be vulnerable to the effects of acid precipitation.

Aside from the protection some populations may receive from living within a national park, Santeetlah Dusky Salamanders are not listed in the U.S. under any state or federal regulations.

ORIGINAL ACCOUNT by Carlos D. Camp and Stephen G. Tilley; photograph by Mike Redmer.

Desmognathus welteri Barbour, 1950

Black Mountain Salamander

Black Mountain Salamanders occur at elevations from 300 to 800 m in the Cumberland Mountains and Cumberland Plateau of southwestern Virginia, southern West Virginia, eastern Kentucky, and north central Tennessee.

Black Mountain Salamanders are found along permanent, small- to medium-sized streams with moderate to steep gradients located in moist forests. Although generally aquatic, Black Mountain Salamanders are also commonly found on the stream banks and adults are frequently associated with large rocks.

The habitats of Black Mountain Salamanders have been mined extensively for coal, with considerable negative effects. Siltation of streams increases dramatically downstream of mining activity, and mountain-top removal, with the resultant deposition of the overburden in the valley below, completely eliminates habitats for Black Mountain Salamanders. Such habitat alterations, as well as the widespread use of Black Mountain Salamanders as fish bait, have resulted in local population declines.

Black Mountain Salamanders are listed as Rare in West Virginia.

ORIGINAL ACCOUNT by J. Eric Juterbock and Zachary I. Felix; photograph by Michael Graziano.

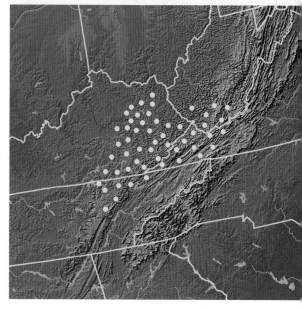

Black Mountain Salamander

Desmognathus wrighti King, 1937

Pigmy Salamander

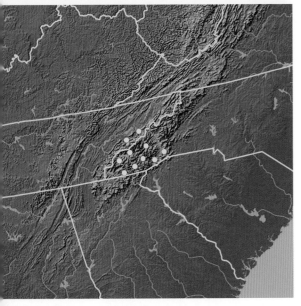

Pigmy Salamander

Pigmy Salamanders occur in the Blue Ridge region of southwestern North Carolina and adjacent eastern Tennessee south and west of the French Broad River at elevations ranging from approximately 750 to 2000 m.

Pigmy Salamanders are especially characteristic of spruce–fir forests though at lower elevations, they will also occur in moist forests dominated by hardwoods. They are highly terrestrial and will hide beneath small logs and rocks or under the bark of trees or stumps. In late fall, they tend to retreat into underground seepage areas. Pigmy Salamanders are active at night and may climb on vegetation up to 1 m above the ground.

In North Carolina, Pigmy Salamanders have been placed on a "Watch List" as species facing increasing amounts of threats to its habitat. In Tennessee, Pigmy Salamanders are considered to be "Very Rare and Imperiled" and in need of management.

ORIGINAL ACCOUNT by Julian R. Harrison; photograph by Mike Redmer.

Ensatina eschscholtzii Gray, 1850

Ensatina

Ensatinas are found on the central coast of British Columbia as far north as the Kitlope Valley. They also occur on Vancouver Island and range south along western Washington and western Oregon to encircle the Central Valley of California and continue south along the coast into northern Baja California (not shown). Seven morphologically distinct forms of variously blotched, spotted or plain-colored Ensatinas are distributed parapatrically in the ring around the California's Central Valley.

Ensatinas inhabit coniferous forest, deciduous forest, oak woodland, coastal sage scrub, and chaparral. Because they require moist, stable microhabitats, they will be found under logs, bark, and moss; under leaf litter; in talus; or in animal burrows. Edge habitats appear to support the highest abundances of Ensatinas. They also may be more abundant on flat or gently sloping shelves above flood level rather than on steep terrain.

Ensatinas usually are common wherever they are present. In Douglas fir forests, they are more abundant in old-growth stands than in young or mature regenerating stands and are consistently more abundant in drier sites. Surface activity by Ensatinas is highly correlated with surface moisture, and they are not commonly encountered above ground during the summer dry season.

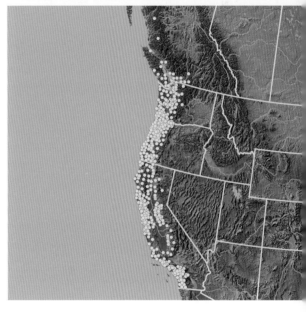

Ensatina

Two of the subspecies of Ensatinas in southern California are listed as California Species of Special Concern.

ORIGINAL ACCOUNT by Shawn R. Kuchta and Duncan Parks; photograph by Dante Fenolio.

Eurycea aquatica (Rose and Bush, 1963)

Brown-backed Salamander

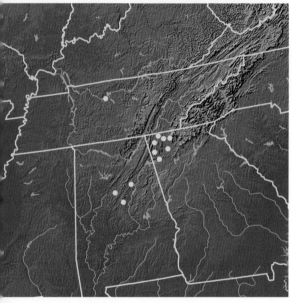

Brown-backed Salamander

Brown-backed Salamanders are known from small springs in the Appalachian Valley and Ridge region of central Alabama and northwestern Georgia, though they appear to be absent south of the Coosa River despite the apparent availability of suitable spring habitats. They have also been reported from the Interior Low Plateaus of Tennessee.

The primary habitats of Brown-backed Salamanders are limestone and dolomitic springs in wooded areas and the streams that issue from them. Brown-backed Salamander larvae generally are nocturnal and appear to prefer the bottoms of shallow, quiet pools that form below riffle areas in rocky streams.

The springs inhabited by Brown-backed Salamanders are very small, making them inherently vulnerable to disturbance.

Brown-backed Salamanders are not listed in the U.S. under any state or federal laws or regulations.

ORIGINAL ACCOUNT by Kenneth H. Kozak and Michael J. Lannoo; photograph by Pierson Hill.

Eurycea bislineata (Green, 1818)

Northern Two-lined Salamander

Northern Two-lined Salamanders are the well-known "yellow salamanders" of the northeastern U.S. and eastern Canada. Their distribution is from eastern Ontario, central Québec, and Labrador, south through New Brunswick and New England to northern Virginia, and west through eastern Ohio and the Kanawha River valley of West Virginia. The northern distributional limits of Northern Two-lined Salamanders in Canada are unclear, and their range may be expanding.

The primary habitats of Northern Two-lined Salamanders include unpolluted bogs, springs, streams, or lakes in wooded areas, provided they lack large predatory fish. Most small streams in the wooded, mountainous areas of their range are rocky, which provides Northern Two-lined Salamanders with a good substrate for nesting. Occasionally, they are found along larger streams and rivers but predation by fish likely limits their extensive exploitation of rocky areas along large streams. Newly metamorphosed juvenile Northern Two-lined Salamanders may migrate into terrestrial habitats up to 100 m from streams.

In general, Northern two-lined Salamanders are widespread and suitable habitat is available across their range.

Northern two-lined Salamanders are listed as Protected by the State of New Jersey.

Northern Two-lined Salamander

ORIGINAL ACCOUNT by David M. Sever; photograph by William Leonard.

Eurycea chamberlaini, Harrison and Guttman, 2003

Chamberlain's Dwarf Salamander

Chamberlain's Dwarf Salamanders are known from scattered localities in the Piedmont and Atlantic Coastal Plain from North Carolina through to southern Alabama.

Chamberlain's Dwarf Salamanders are semi-terrestrial and normally occupy the margins of streams or seepages, or floodplain or pond sites.

Chamberlain's Dwarf Salamanders have no legal protection in the U.S., but with the possible and likely exception of some local populations, are probably not in any immediate jeopardy.

ORIGINAL ACCOUNT by Julian R. Harrison; photograph by Pierson Hill.

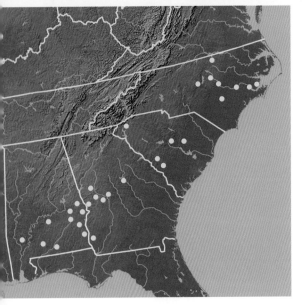

Chamberlain's Dwarf Salamander

Eurycea chisholmensis Chippindale, Price, Wiens, and Hillis, 2000

Salado Salamander

Salado Salamanders inhabit springs associated with the Balcones Escarpment near Salado Creek at 183 m elevation in east central Texas.

Salado Salamanders are completely aquatic and are known only from the immediate vicinity of spring outflows, where they hide under rocks and in gravel. The water temperature in the springs off the Edwards Plateau is relatively constant throughout the year and typically ranges from 18 to 20°C.

Most of the spring outlets at Salado Creek have been modified over the past 150 years, and groundwater contamination incidents such have occurred in the recent past may pose a threat to Salado Salamanders.

Salado Salamanders are a Candidate species for U.S. federal listing.

ORIGINAL ACCOUNT by Paul T. Chippindale; photograph by Richard D. Bartlett.

Salado Salamander

Eurycea cirrigera (Green, 1830)

Southern Two-lined Salamander

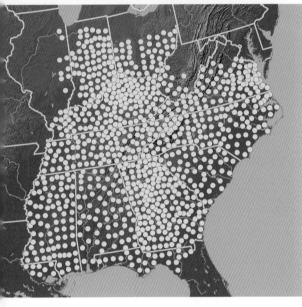

Southern Two-lined Salamander

The range of Southern Two-lined Salamanders is generally to the east of the Wabash and Mississippi rivers from Illinois and Indiana south to the Gulf Coast, excluding peninsular Florida. Their northern range limit is across northern Indiana and Ohio, through western West Virginia, and across to central Virginia.

Southern Two-lined Salamanders are semi-aquatic and can be found in such habitats as streams, pools, seeps, ditches, damp woods, and the heads of ravines throughout their range. Southern Two-lined Salamanders frequently use stream habitats with coarse sand and gravel, as well as broken limestone rock, leaf litter, and crayfish burrows. Juveniles are found under stones and other cover objects at the edges of aquatic habitats.

Southern Two-lined Salamanders are abundant throughout most of their range. They are reasonably tolerant of water polluted with sewage or other organic matter. As with most forest salamanders, Southern Two-lined Salamanders are vulnerable to habitat destruction through activities such as clear-cutting and habitat degradation due to acid mine drainage and acid deposition.

Southern Two-lined Salamanders receive no protection in the U.S. under any state or federal laws or regulations.

ORIGINAL ACCOUNT by Thomas K. Pauley and Mark B. Watson; photograph by Dante Fenolio.

Eurycea guttolineata (Holbrook, 1838)

Three-lined Salamander

Three-lined Salamanders are distributed throughout much of the southeastern U.S. east of the Mississippi River from Louisiana north into western Kentucky and east to the Atlantic seaboard, avoiding both the highlands of the Appalachians and the lowlands of the Florida peninsula. Populations of Three-lined Salamanders are almost always found below elevations of 1000 m.

Three-lined Salamanders are mainly terrestrial as adults, though they rarely are found considerable distances from wetlands and are good swimmers, at home in the water. They are most abundant in river-bottom wetlands and in the vicinity of springs and streams where seepage keeps the ground moist. Three-lined Salamanders are primarily nocturnal and may be found during the day under cover objects. Surface activity is closely tied with surface moisture. Adult Three-lined Salamanders are most likely to be encountered foraging on humid or rainy nights shortly after sunset.

Three-lined Salamanders remain abundant throughout much of their range. While the loss of bottomland hardwood forests throughout the Southeast of the U.S. has undoubtedly resulted in the extirpation of many populations, direct links between habitat loss and population declines have not been demonstrated.

Three-lined Salamanders are not listed in the U.S. under any state or federal regulations.

ORIGINAL ACCOUNT by Travis J. Ryan and Brooke A. Douthitt; photograph by Brad Moon.

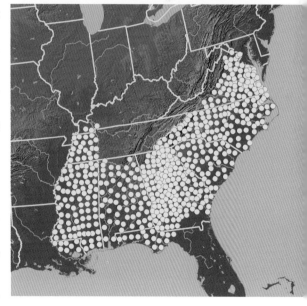

Three-lined Salamander

Eurycea junaluska (Sever, Dundee and Sullivan, 1976)

Junaluska Salamander

Junaluska Salamanders are known only from in and near the Great Smoky Mountains of eastern Tennessee and western North Carolina in tributaries of the Little Tennessee River and the Little Pigeon River at elevations between about 360 and 610 m.

Junaluska Salamanders are active during warm, rainy nights during early spring near creeks, on streambanks and in streams. In colder months, adult Junaluska Salamanders are most commonly found within streams, whereas in warmer months, they are found predominantly in the forests surrounding the streams. Within the streams, both adult and larval Junaluska Salamanders hide beneath large flat rocks in regions of shallow water and moderate stream flow.

Tennessee populations of Junaluska Salamanders appear stable. In North Carolina, however, they were rare to begin with and remain so.

In North Carolina, Junaluska Salamanders are currently listed as a Species of Special Concern but they have been proposed for Threatened status.

ORIGINAL ACCOUNT by Travis J. Ryan and David M. Sever; photograph by Nathan Haislip.

Junaluska Salamander

Eurycea latitans Smith and Potter, 1946

Cascade Caverns Salamander

Cascade Caverns Salamanders are found only in the Cascade Caverns of the Balcones Escarpment of south central Texas. This population includes individuals with a spectrum of morphologic features ranging from highly cave-associated morphologies to surface-like morphologies.

Cascade Caverns Salamanders are completely aquatic and thus are known only from caves that contain water or from the immediate vicinity of spring outflows. Individuals in caves are often seen in the open on submerged rocks or mud, whereas individuals from springs hide under rocks and leaves and in gravel substrate.

Cascade Caverns Salamanders can be common at spring outflows but their distribution appears to be limited.

Cascade Caverns Salamanders are listed as Threatened in Texas, but have no special recognition by the U.S. federal government.

ORIGINAL ACCOUNT by Paul T. Chippindale; photograph by Dante Fenolio.

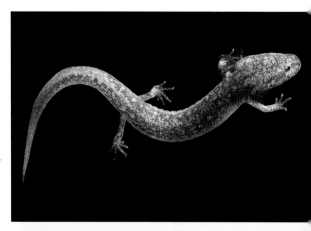

Cascade Caverns Salamander

Eurycea longicauda (Green, 1818)

Long-tailed Salamander

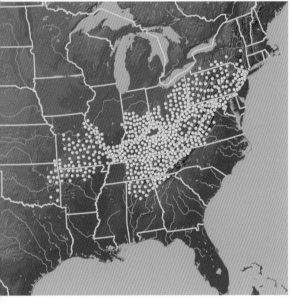

Long-tailed Salamander

Long-tailed Salamanders are distributed throughout the Ozark Highlands, the Appalachian Highlands, and the Ohio River Valley. There is a narrow connection between the Ozarks and the rest of their range through southern Illinois and western Kentucky.

Long-tailed Salamanders are mainly terrestrial and are found in and beneath old rotting logs and under stones. They are commonly found in crevices of shale and beneath stones and rock fragments near the margins of streams. Adult Long-tailed Salamanders freely enter water and swim with ease; they will also inhabit caves. Long-tailed Salamanders feed on humid and rainy nights, where they are most active during the first few hours after dark. Postmetamorphic juveniles can be abundant near pond edges immediately following metamorphosis, taking refuge under rocks, fallen tree trunks, and even beneath tree bark. In general, juvenile Long-tailed Salamanders are found closer to the water than are adults.

Long-tailed Salamanders can be locally abundant, but populations have undoubtedly been lost because of habitat loss, effects of coal mining, and clear-cutting. However, there are no robust distributional studies to document changes in their distribution.

Long-tailed Salamanders are listed as Threatened in Kansas and New Jersey, and a Species of Special Concern in North Carolina.

ORIGINAL ACCOUNT by Travis J. Ryan and Christopher Conner; photograph by Dante Fenolio.

Eurycea lucifuga Rafinesque, 1822

Cave Salamander

Cave Salamanders are found in the Ozark Plateau and Ouachita Mountains of southern Missouri, northern Arkansas and nearby Oklahoma and Kansas, across southern Illinois, in the interior Low Plateau region from the Mitchell and Muskatatuck Plateaus in southern Indiana and Ohio south through the Pennyrile region of Kentucky and the Cumberland Plateau to northern Alabama and up the Appalachian Valley and Ridge region from northwestern Georgia to northern Virginia.

Cave Salamanders are terrestrial and are associated with limestone regions where they live most abundantly, climbing over walls and ledges in the semidarkness of the twilight regions of caves. However, their predilection for cave habitats is probably overemphasized, as they are also found under stones, logs, and other surface matter outside of caves, forested limestone ravines, or even spring-fed cypress swamps, located away from rock bluffs.

Cave Salamanders are dependent upon, or associated with, caves and similar limestone features. Although the biggest threat to Cave Salamanders and other cave-dwelling animals may be their extremely localized occurrence, actions that directly degrade subsurface habitat or surface terrestrial and/or aquatic habitats will also negatively affect populations.

Cave Salamanders are listed as Endangered in Ohio, Mississippi, and Kansas, and considered Rare in West Virginia.

ORIGINAL ACCOUNT by J. Eric Juterbock; photograph by Dante Fenolio.

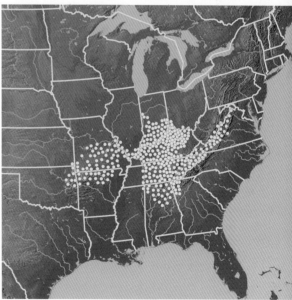

Cave Salamander

Eurycea multiplicata (Cope, 1869)

Many-ribbed Salamander

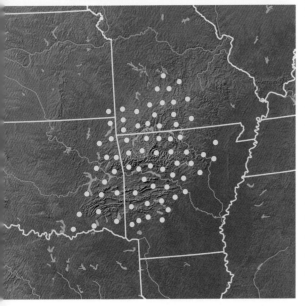

Many-ribbed Salamander

Many-ribbed Salamanders occur in the Ozark Plateau and in the Boston and Ouachita Mountains and associated lowland rocky formations in extreme southeastern Kansas, southwestern Missouri, eastern Oklahoma, and west central Arkansas at elevations between 110 and 760 m.

Many-ribbed Salamanders are found in cool, moist habitats near water in association with springs and permanent streams and within the wetter portions of such streams during drier periods. They tend to be found under stones, logs, and other large materials in streams and springs, both in the open and in the twilight zone of caves. During daylight, larval Many-ribbed Salamanders are usually found under stones in slow-moving streams, and they will inhabit more ephemeral portions of streams than will adults. Gilled, neotenic adults are common among Many-ribbed Salamanders that inhabit caves or streams draining caves.

Many-ribbed Salamanders inhabit an extensive region of rugged, hilly, and mountainous terrain, most of which is not amenable to agriculture or developing urban settings. Despite considerable logging over much of the terrain and use of flatter areas for cattle grazing, substantial populations of Many-ribbed Salamanders are found in many settings, even within properties that people have developed. The only threats to the species are an increasing number of recreational homes scattered throughout the region and some conversion of caves into commercial attractions.

Many-ribbed Salamanders are not listed in the U.S. under any state or federal laws or regulations.

ORIGINAL ACCOUNT by Stanley E. Trauth and Harold A. Dundee; photograph by John Clare.

Eurycea nana Bishop, 1941

San Marcos Salamander

San Marcos Salamanders are found only in out-flows of San Marcos Springs in central Texas at an elevation of 174 m.

San Marcos Salamanders are neotenic, completely aquatic and extremely abundant within their severely limited range. Though San Marcos Springs has been heavily modified in the past century to form a small lake, San Marcos Salamanders occur throughout much of this lake and extend about 150 m into the most upstream portion of the San Marcos River. They are found in mats of blue-green algae, under rocks and in gravel substrate at water depths down to several meters.

San Marcos Salamanders are listed as Threatened both by the State of Texas and the U.S. federal government.

ORIGINAL ACCOUNT by Paul T. Chippindale and Joe N. Fries; photograph by Gary Nafis.

San Marcos Salamander

Eurycea naufragia Chippindale, Price, Wiens and Hillis, 2000

Georgetown Salamander

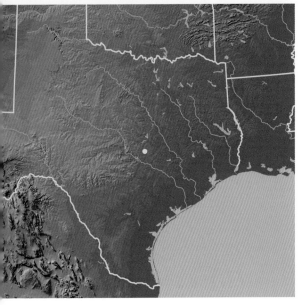

Georgetown Salamander

Georgetown Salamanders occur in springs of the San Gabriel River drainage at an elevation of 230 m in the vicinity of Georgetown, in central Texas.

Georgetown Salamanders are known only from the immediate vicinity of spring outflows and from two water-containing caves, where they live under rocks and leaves and in gravel substrate. They are neotenic and completely aquatic throughout life.

Several populations of Georgetown Salamanders occurred adjacent to Lake Georgetown, and it is likely that they were submerged when that man-made lake was created. Springs in Georgetown's San Gabriel Park have been heavily modified, and the continued existence of Georgetown Salamanders at these springs appears precarious.

Georgetown Salamanders currently are considered as a Candidate species for U.S. federal listing but they have not been listed by the State of Texas.

ORIGINAL ACCOUNT by Paul T. Chippindale; photograph by Richard D. Bartlett.

Eurycea neotenes Bishop and Wright, 1937

Texas Salamander

Texas Salamanders live in springs and caves in the Edwards Plateau of south central Texas.

Texas Salamanders are neotenic, and natural metamorphosis is unknown. They are completely aquatic and reside under rocks and leaves or in gravel under the water. Texas Salamanders appear to retreat into subterranean crannies and fissures during dry conditions, allowing them to survive temporary drying of surface springs.

Texas Salamanders may be abundant at the spring outflows where they occur, but their distribution appears to be limited and patchy. Despite this, they receive no protection by either the State of Texas or the U.S. government.

ORIGINAL ACCOUNT by Paul T. Chippindale; photograph by Richard D. Bartlett.

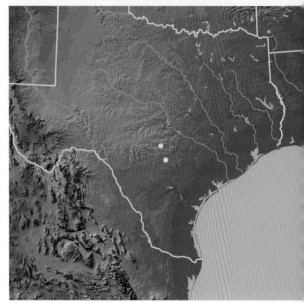

Texas Salamander

Eurycea pterophila Burger, Smith, and Potter, 1950

Fern Bank Salamander

Fern Bank Salamanders have a limited and patchy distribution in the Blanco River drainage off the Edwards Plateau in central Texas.

Fern Bank Salamanders may be common at spring outflows and in caves. They are entirely aquatic and neotenic as they retain their external gills and larval form throughout life. They live in the immediate vicinity of spring outflows, under rocks and leaves, and in gravel substrate and eat mainly small aquatic invertebrates.

Fern Bank Salamanders are not listed by either the State of Texas or the U.S. government.

ORIGINAL ACCOUNT by Paul T. Chippindale; photograph by Dante Fenolio.

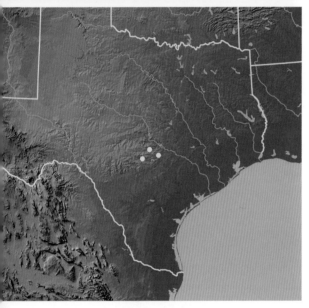

Fern Bank Salamander

Eurycea quadridigitata, Holbrook, 1842

Dwarf Salamander

Dwarf Salamanders occur primarily on the Atlantic and Gulf Coastal Plains from northern North Carolina south throughout much of peninsular Florida, west to the Brazos River of eastern Texas, and north into southern Arkansas.

Dwarf Salamanders are found beneath cover objects at the edges of ponds and swamps as well as in seeps and among leaf litter in springs. They breed in such places as temporary ponds, Carolina bays, and hammock ponds, though they may also use small streams. Adult Dwarf Salamanders avoid predation and desiccation during their migrations to and from their breeding sites by traveling beneath the leaf litter.

Dwarf Salamanders currently are given no special protective status at either the state or federal level in the U.S.

ORIGINAL ACCOUNT by Ronald M. Bonett and Paul T. Chippindale; photograph by Brad Moon.

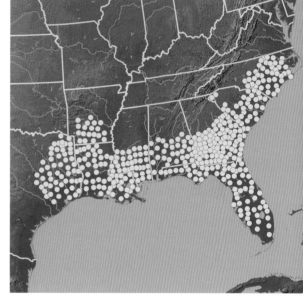

Dwarf Salamander

Eurycea rathbuni (Stejneger, 1896)

Texas Blind Salamander

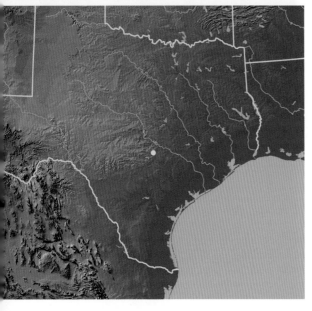

Texas Blind Salamander

Texas Blind Salamanders are known from several caves, wells, and pipes that intersect the San Marcos Pool of the Edwards Aquifer at an elevation of about 188 m in San Marcos, Texas.

Texas Blind Salamanders are completely aquatic and wholly subterranean. In caverns of the San Marcos Pool of the Edwards Aquifer, they have been observed climbing rock surfaces or swimming in open water. Texas Blind Salamanders can reliably be observed at a tiny cave opening into the Edwards Aquifer in San Marcos, Texas. When the cave floods, individuals are active on the surface, even in broad daylight. Natural metamorphosis is unknown in Texas Blind Salamanders

Populations of Texas Blind Salamanders have been lost, and they are listed as Endangered by both the State of Texas and the U.S. federal government.

ORIGINAL ACCOUNT by Paul T. Chippindale; photograph by Dante Fenolio.

Eurycea robusta (Longley, 1978)

Blanco Blind Salamander

Blanco Blind Salamanders are known only from beneath the Blanco River just east of San Marcos, Texas, on the Edwards Plateau. A single specimen was collected in 1951, and no further specimens have been collected since.

Blanco Blind Salamanders are neotenic, completely subterranean, and wholly aquatic. Virtually nothing else is known about them.

A 1995 petition to list Blanco Blind Salamanders as U.S. federally Endangered was rejected because of lack of information on its status and distribution. However, the State of Texas lists them as Threatened.

ORIGINAL ACCOUNT by Paul T. Chippindale.

Blanco Blind Salamander

Eurycea sosorum Chippindale, Price, and Hillis, 1993

Barton Springs Salamander

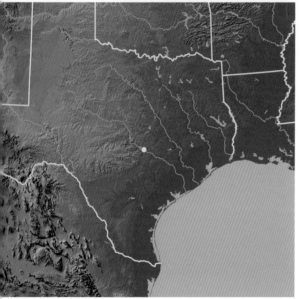

Barton Springs Salamander

Barton Springs Salamanders inhabit only Barton Springs pool, a spring-fed swimming hole in the city of Austin, Texas, and three other springs that are immediately adjacent. These spring outflows are fed by the Barton Springs segment of the Edwards Aquifer of central Texas.

Barton Springs Salamanders are completely aquatic and appear to dwell primarily in surface waters at depths ranging from a few cm to about 5 m, though they may also use subterranean habitats. Larvae, juveniles, and adults all live amid cobbles, gravel, and aquatic plants at the bottom of the water. Barton Springs Salamanders are neotenic, and natural metamorphosis is unknown. They prey mainly on small aquatic invertebrates, especially amphipods.

Barton Springs Salamanders underwent a major decline in abundance probably due, in part, to cleaning procedures used by the City of Austin at Barton Springs Pool. These pool-maintenance procedures have since been modified to balance the need for pool maintenance with protection of Barton Springs Salamanders.

Barton Springs Salamanders are listed as an Endangered species in the U.S.

ORIGINAL ACCOUNT by Paul T. Chippindale and Robert Hansen; photograph by Dante Fenolio.

Eurycea spelaea Stejneger, 1893

Grotto Salamander

The distribution of Grotto Salamanders spans the Salem and Springfield regions of the Ozark uplift in Missouri, Oklahoma, extreme southeastern Kansas, and Arkansas.

Grotto Salamanders are limited to limestone caves and underground passages in the karst formations of the Ozark Plateau and are most frequently found beyond the twilight zone of caves on moist rock walls. The juvenile stage is short, and sexual maturity appears to occur at, or shortly after, metamorphosis. Small and intermediate-sized larvae have functional eyes that degenerate in older larvae and adults. Eyelids may grow over vestigial eyes in some adults.

Detailed records of sightings of Grotto Salamanders dating as far back as the 1930s, combined with recent surveys, have demonstrated that populations of Grotto Salamanders have been stable for over 25 years. Larval Grotto Salamanders are sensitive to flooding of subterranean streams and are consumed by predatory fishes. Habitat alteration, though, poses the greatest threat to the survival of Grotto Salamanders, which seem particularly sensitive to disturbance or impurities introduced into subterranean waterways and caves.

Grotto Salamanders are listed in three of the four states where they are found. Arkansas is taking steps to list them as well.

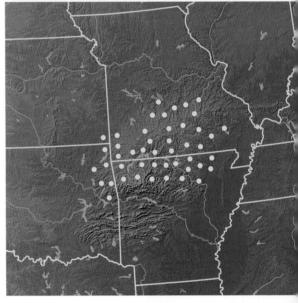

Grotto Salamander

ORIGINAL ACCOUNT by Dante B. Fenolio and Stanley E. Trauth; photograph by Michael Graziano.

Eurycea tonkawae Chippindale, Price, Wiens, and Hillis, 2000

Jollyville Plateau Salamander

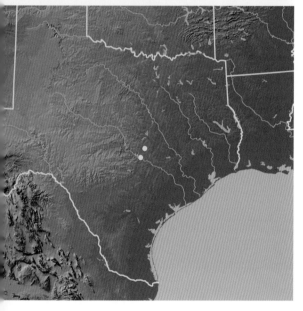

Jollyville Plateau Salamander

Jollyville Plateau Salamanders are known only from springs and caves at the margin of the Jollyville Plateau near Brushy Creek in the city of Austin, Texas.

Jollyville Plateau Salamanders are completely aquatic and known only from the vicinity of spring outflows, particularly in areas of cobbled substrate, in which they hide among leaves and gravel.

Jollyville Plateau Salamanders were discovered and described only recently, so little is known of their historical abundance, though they may be common at some spring outflows. Urban and suburban development is known to have negatively affected some populations.

Jollyville Plateau Salamanders are federally listed as Threatened under the U.S. Endangered Species Act.

ORIGINAL ACCOUNT by Paul T. Chippindale; photograph by Richard D. Bartlett.

Eurycea tridentifera Mitchell and Reddell, 1965

Comal Blind Salamander

The distribution of Comal Blind Salamanders includes several caves in the Cibolo Sinkhole Plain of the southeastern Edwards Plateau region of south central Texas.

Comal Blind Salamanders are found on rock and mud substrates in underground streams and pools deep within caves. They are neotenic and thus retain their gilled, larval form through life.

It is difficult to assessment how abundant Comal Blind Salamanders may be, though they appear to be increasingly scarce.

Comal Blind Salamanders are listed as Threatened by the State of Texas, yet they have not attracted federal attention.

ORIGINAL ACCOUNT by Paul T. Chippindale; photograph by Dante Fenolio.

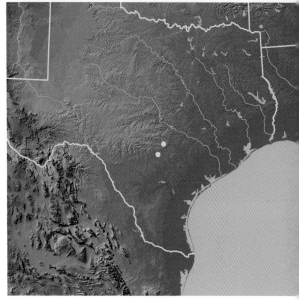

Comal Blind Salamander

Eurycea troglodytes Baker, 1957

Valdina Farms Salamander

Valdina Farms Salamanders are known from a large and wide-ranging array of springs and caves, including the Valdina Farms sinkhole complex, in the Edwards Plateau region of south central Texas.

Valdina Farms Salamanders are known only from caves that contain water and from the immediate vicinity of spring outflows. In most populations of Valdina Farms Salamanders, metamorphosis is unknown and they are completely aquatic and neotenic. However, natural metamorphosis has been observed in Valdina Farms Salamanders from several springs in the Sabinal River drainage, and remains of terrestrial invertebrates have been found in the stomachs of some transformed individuals that were captured in the water, suggesting that they may venture short distances onto land.

Construction of a diversion dam temporarily submerged the Valdina Farms sinkhole and allowed catfish and other predators to enter. Since 1987, Valdina Farms Salamanders have not been found at that site, even in areas of the cave where they once were common.

Valdina Farms Salamanders have not been given special conservation status by either the State of Texas or the U.S. government.

ORIGINAL ACCOUNT by Paul T. Chippindale; Photograph by Dante Fenolio.

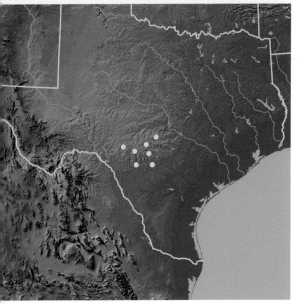

Valdina Farms Salamander

Eurycea tynerensis Moore and Hughes, 1939

Oklahoma Salamander

Oklahoma Salamanders are found on the Spring-field Plateau of northwestern Arkansas, north-eastern Oklahoma, and southwestern and south central Missouri.

Oklahoma Salamanders are permanently aquatic and neotenic. They tend to inhabit subter-ranean aquatic habitats that range from cool, clear, swift streams that contain coarse gravel, in which they seek cover, to shallow, slowly moving streams containing medium-sized rocks that are only par-tially embedded, to small springs and seeps with moist leaf litter over a mud-and-detritus substrate. Oklahoma Salamanders may be most abundant in areas where aquatic invertebrate densities are high, either because they prefer areas of high prey density or because they and aquatic invertebrates have the same preferences for water conditions or means of avoiding predators.

Oklahoma Salamanders were once listed as Rare on Missouri's Rare and Endangered Species List, but are no longer recognized as a valid taxon there. In Oklahoma, they are provided with no special listed status. In Arkansas, Oklahoma Salamanders are considered to be a Species of Special Concern and collecting permit requests are closely monitored.

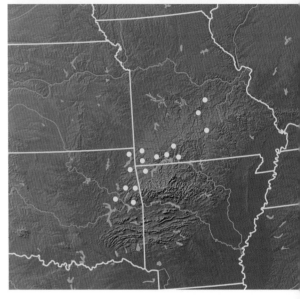

Oklahoma Salamander

ORIGINAL ACCOUNT by Ronald M. Bonett; photograph by John Clare.

Eurycea waterlooensis Hillis, Chamberlain, Wilcox, and Chippindale, 2001

Austin Blind Salamander

Austin Blind Salamanders are known only from the outflows of Barton Springs in the city of Austin, Texas.

Austin Blind Salamanders are completely aquatic and are known only from juveniles that probably washed out accidentally from spring outflows. They are almost certainly cave dwellers.

Austin Blind Salamanders are federally listed in the U.S. as Endangered.

ORIGINAL ACCOUNT by Paul T. Chippindale; photograph by Dante Fenolio.

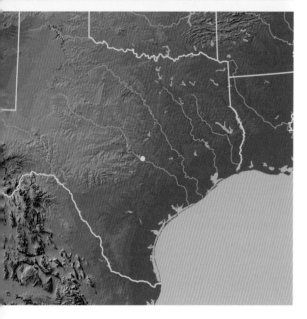

Austin Blind Salamander

Eurycea wilderae Dunn, 1920

Blue Ridge Two-lined Salamander

Blue Ridge Two-lined Salamanders inhabit the whole of the southern Blue Ridge region, including eastern Virginia, North Carolina, South Carolina, eastern Tennessee, and northeastern Georgia. They are found from base-level streams to the tops of the highest peaks at elevations of about 1900 m.

Blue Ridge Two-lined Salamanders occur naturally in every rocky mountain stream in the southern Blue Ridge. They often are found under rocks, logs, leaves, and other detritus in streamside habitats. Male Blue Ridge Two-lined Salamanders frequently are found with females under rocks in streams during the spring mating period. Females subsequently stay in the water with their nests, whereas the males move into terrestrial habitats.

Habitats for Blue Ridge Two-lined Salamanders remain plentiful throughout their range, and animals remain abundant in suitable streams. They are still likely to be found in every stream that has not been damaged by pollution, siltation, deforestation, channeling, or other insults. Indeed, they appear to be rather resilient and can still commonly occur even in seemingly inhospitable streams that have become murky and exposed as businesses and homes alongside them have flourished.

There are no current conservation concerns associated with Blue Ridge Two-lined Salamanders, and they are not listed in the U.S. under any state or federal regulations.

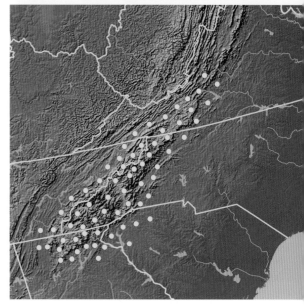

Blue Ridge Two-lined Salamander

ORIGINAL ACCOUNT by David M. Sever; photograph by Michael Graziano.

Gyrinophilus gulolineatus Brandon, 1965

Berry Cave Salamander

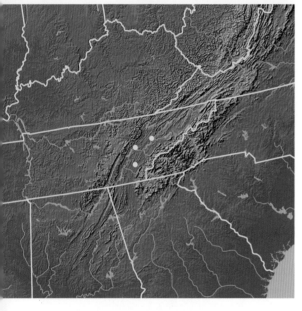

Berry Cave Salamanders are known only from the Appalachian Valley and Ridge region west of the Great Smoky Mountains in eastern Tennessee.

Berry Cave Salamanders are cave dwellers associated with inflow or sinkhole caves, which provide a detritus base that appears to be necessary for them to live in.

Populations of Berry Cave Salamanders are declining because of aboveground habitat destruction and subsequent effects on water quality within caves.

Berry Cave Salamanders are designated by the U.S. Fish and Wildlife Service as a candidate species for listing under the Endangered Species Act.

ORIGINAL ACCOUNT by Christopher K. Beachy; photograph by Brad Glorioso.

Berry Cave Salamander

Gyrinophilus palleucus McCrady, 1954

Tennessee Cave Salamander

Tennessee Cave Salamanders are associated with limestone cave systems in eastern and central Tennessee, northern Alabama, and northwestern Georgia.

Tennessee Cave Salamanders are found in sinkhole-type caves or phreatic cave systems in the vicinity of sinkholes. The nutrients that flow into these cave systems support an abundance of invertebrates, which form the prey base for the salamanders. Within these caves, Tennessee Cave Salamanders are found under rocks in rocky and sandy substrates in quiet, shallow pools.

The distribution of Tennessee Cave Salamanders has been affected by the indirect effects of deforestation on the surface, and most populations appear to be declining.

Tennessee Cave Salamanders are listed a Species of Special Concern in Georgia and as a Protected Species in Alabama, whereas the Tennessee Wildlife Resources Agency lists them as Threatened. Although the U.S. Fish and Wildlife Service had listed Tennessee Cave Salamanders as a Category 2 candidate for federal listing in 1994, they were not included in any subsequent federal list.

ORIGINAL ACCOUNT by Christopher K. Beachy; photograph by Brad Glorioso.

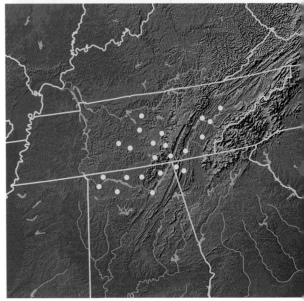

Tennessee Cave Salamander

Gyrinophilus porphyriticus (Green, 1827)

Spring Salamander

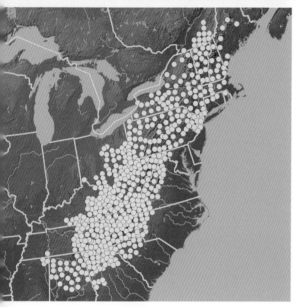

Spring Salamander

Spring Salamanders range from Maine and southern Québec southwest along the Appalachian Mountains and associated foothills through to northern Georgia, central Alabama, and adjacent Mississippi.

Spring Salamanders are most abundant in the headwater sections of small streams that lack fishes and in seepages and caves; they can sometimes be found in roadside ditches. In the Piedmont of South Carolina, for example, populations of Spring Salamanders are associated with springs and small streams in deep ravines covered with mature hardwood forest. Spring Salamanders tend to be more common in caves than in springs in limestone areas of the south central Appalachians. Spring Salamanders are notoriously difficult to find, as they tend to live in habitats that are difficult to penetrate and, for the most part, are only occasionally active on the surface.

Deforestation is a threat to many populations of Spring Salamanders.

Spring Salamanders are considered Endangered in Mississippi, Threatened in Connecticut, of Special Concern in Massachusetts, and of Concern in Rhode Island. In Canada federally, they are listed as a species of Special Concern and, provincially, as Extirpated in Ontario and *"vulnérable"* in Québec.

ORIGINAL ACCOUNT by Christopher K. Beachy; photograph by Michael Graziano.

Gyrinophilus subterraneus Besharse and Holsinger, 1977

West Virginia Spring Salamander

West Virginia Spring Salamanders are known only from General Davis Cave at an elevation of 503 m in southeastern West Virginia.

West Virginia Spring Salamanders live on the muddy banks along the stream that flows inside General Davis Cave and into the short ravine in the vicinity of the cave mouth. The stream usually varies in depth from 15 to 30 cm, but will flood following rains. The banks of the stream are muddy and steep and contain a large amount of decaying leaf litter that is occasionally washed into the cave by floods. This leaf litter offers a source of nutrients for the cave's invertebrates that are eaten by the salamanders. West Virginia Spring Salamanders have been observed almost 2 km into General Davis Cave.

Because of their restricted distribution, West Virginia Spring Salamanders are on West Virginia's Rare species list.

ORIGINAL ACCOUNT by Christopher K. Beachy; photograph by Dante Fenolio.

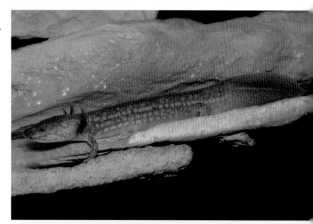

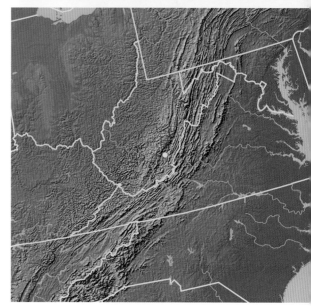

West Virginia Spring Salamander

Haideotriton wallacei Carr, 1939

Georgia Blind Salamander

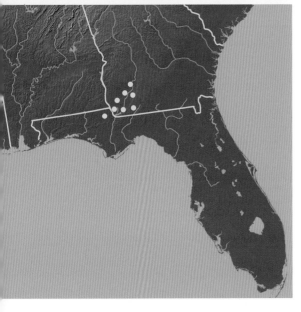

Georgia Blind Salamander

Georgia Blind Salamanders inhabit the Floridan Aquifer where it circulates in underground passageways in limestone beneath the Marianna Lowlands of western Florida and the Dougherty Plain of southwestern Georgia.

Georgia Blind Salamanders are found in pools and underground streams, especially in caves where bats defecate over or near the water. Individuals move about slowly, resting on bottom sediments or climbing on limestone sidewalls and ledges underwater. Georgia Blind Salamanders are thought to be neotenic, as no transformed animals have been collected and their larvae show no response to metamorphic agents.

At least two caves in which Georgia Blind Salamanders were known to occur in the Marianna Lowlands have been destroyed by human activities, but in undisturbed caves populations appear stable. Nevertheless, agricultural drawdowns of the Floridan Aquifer and pollution from agricultural runoff may be detrimental to Georgia Blind Salamanders.

Of all the caves inhabited by Georgia Blind Salamanders, only those in the Marianna Caverns State Park in Florida are specifically listed and protected.

ORIGINAL ACCOUNT by D. Bruce Means; photograph by Pierson Hill.

Hemidactylium scutatum Temminck and Schlegel, 1838

Four-toed Salamander

The main part of the range of Four-toed Salamanders extends from northeast Minnesota east through Upper Michigan to southern Ontario and Quebec, south to southern Alabama, and east to the Atlantic seaboard, but this range is highly discontinuous, and disjunct populations are to be found to the east far as Nova Scotia and Maine, to the west all the way to Oklahoma, and to the south to include parts of Florida and Louisiana.

Four-toed Salamanders are found in areas of mature hardwood or coniferous forest, principally associated with ponds, bogs, swamps, or sluggish streams that are free of fish and are near, but not necessarily directly adjacent to, forests. Females nest just above these still waters and brood their eggs until hatching.

The abundance of Four-toed Salamanders appears to be in decline because of draining and loss of vernal ponds and other small wetlands.

Four-toed Salamanders are considered Rare and worthy of Special Concern in Maine and Minnesota. In Québec, they are considered likely to be designated endangered (*menacée*) or threatened (*vulnerable*).

ORIGINAL ACCOUNT by Reid N. Harris; photograph by Todd Pierson.

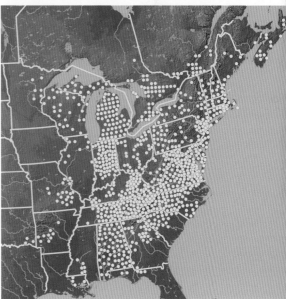

Four-toed Salamander

Hydromantes brunus Gorman, 1954

Limestone Salamander

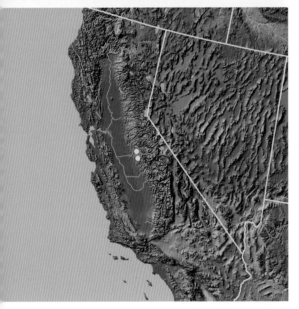

Limestone Salamander

Limestone Salamanders occur only in the western foothills of the Sierra Nevada of California along a short stretch of the Merced River, including a short distance up the North Fork of the Merced River, at elevations of 365 to 760 m.

Limestone Salamanders are found in small areas of moss-covered or barren talus, as well as in rock crevices and even in abandoned mine tunnels. They have been located under slabs of slate and irregularly shaped pieces of limestone. The surrounding vegetation is mainly chaparral, with a scattering of gray pine and with California laurel in more humid sites. Limestone Salamanders have been found on relatively level ground but are more typically encountered on steep slopes, where they use their tails to assist in locomotion.

It is likely that any widening of roads would destroy habitat for Limestone Salamanders, which is already highly restricted.

Limestone Salamanders are listed as a Threatened species by the California Department of Fish and Game.

ORIGINAL ACCOUNT by David B. Wake and Theodore J. Papenfuss; photograph by William Leonard.

Hydromantes platycephalus (Camp, 1916)

Mount Lyell Salamander

Mount Lyell Salamanders are restricted to the Sierra Nevada of California, from the Owens Valley north to the Sonora Pass, with isolated populations occurring further north on the Sierra Buttes, near the northern end of the Sierra Nevada and at Smith Lake in the Desolation Wilderness west of Lake Tahoe. They occur at elevations from 1220 m to about 3600 m.

Mount Lyell Salamanders commonly are found in granite talus slopes downslope from melting snowfields that persist long into, or even through, the summer in the high Sierra Nevada. They favor living under and among flat or roughly quadrangular granite boulders that rest on granite bedrock over which flows a thin film of water. Mount Lyell Salamanders also are encountered under rocks at the edges of streams and, at low elevations, under rocks or, occasionally, pieces of wood in direct contact with moist soil.

Much of the range of Mount Lyell Salamanders lies within national parks and wilderness areas, so there are few threats from human activities. Nevertheless, road construction in the Sonora Pass region could harm the excellent habitat for Mount Lyell Salamanders that occurs near the roadside.

Mount Lyell Salamanders are protected as a Species of Special Concern by the California Department of Fish and Game.

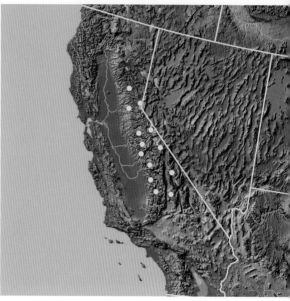

Mount Lyell Salamander

ORIGINAL ACCOUNT by David B. Wake and Theodore J. Papenfuss; photograph by Gary Nafis.

Hydromantes shastae Gorman and Camp, 1953

Shasta Salamander

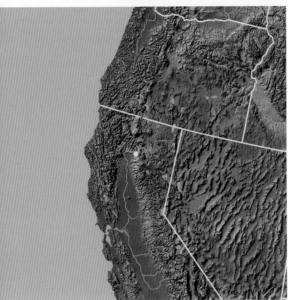

Shasta Salamander

Shasta Salamanders occur at elevations between 300 and 900 m in the vicinity of the Shasta Lake Reservoir, which lies at the confluence of the Sacramento and Pit rivers south of Mt. Shasta in northern California.

Shasta Salamanders mainly occur in areas of limestone outcrops in lightly to densely forested areas dominated by oaks and pines. Though they are generally found in small to large caves, Shasta Salamanders are also commonly encountered at the surface under rocks and logs and in leaf litter from late autumn to early spring. At night, they are active on moist rock faces.

Much of the distribution of Shasta Salamanders is on national forest land; however, a number of populations are small and isolated, and human activities could easily lead to their destruction by relatively small amounts of habitat disturbance.

Shasta Salamanders are considered Threatened by the California Department of Fish and Game, meaning that they may not be taken or possessed at any time.

ORIGINAL ACCOUNT by David B. Wake and Theodore J. Papenfuss; photograph by William Leonard.

Phaeognathus hubrichti Highton, 1961

Red Hills Salamander

Red Hills Salamanders are known from the Red Hills region between the Alabama River and the Conecuh River in south central Alabama.

Red Hills Salamanders are found in steep-sided ravines within the Tallahatta and Hatchetigbee geologic formations, which consist of extremely porous rocks with a large water-holding capacity. Such ravines are humid and have friable soils ideal for Red Hills Salamanders to construct their burrows. The forest canopy in the ravines consists of broad-leaved deciduous trees that provide shade and retain high humidity.

The abundance of Red Hills Salamander may be drastically reduced at some sites depending on historical and current land-use practices, particularly forestry. Virtually all habitat for Red Hills Salamander is on private land, with only a small amount owned by the U.S. Army Corps of Engineers and the State of Alabama.

Red Hills Salamanders are listed as Threatened by the U.S. Endangered Species Act and are listed as a protected nongame species by the State of Alabama.

ORIGINAL ACCOUNT by C. Kenneth Dodd Jr.; photograph by Dante Fenolio.

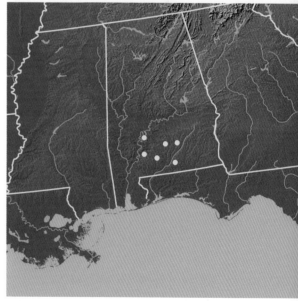

Red Hills Salamander

Plethodon ainsworthi Lazell, 1998

Bay Springs Salamander

Bay Springs Salamanders are known from only two poorly preserved specimens collected near Tallahoma Creek, south of Bay Springs in south central Mississippi.

Bay Springs Salamanders were found in what is now a 2-ha woods of sweet gums, tulip trees, water oaks, white oaks, and loblolly pines.

The conservation status of Bay Springs Salamanders is unknown, and they are possibly extinct. The region where they were found has been planted to pine along the steep slopes, even across the tiny trickling streams in the ravines. The type of locality itself retains a small amount of reasonable forest but is surrounded now by a cow pasture, a clear-cut, and an old farmyard now covered with bamboo.

Bay Springs Salamanders are not listed at either the state or federal level in the U.S.

ORIGINAL ACCOUNT by James Lazell; photograph by Jonathan Woodward.

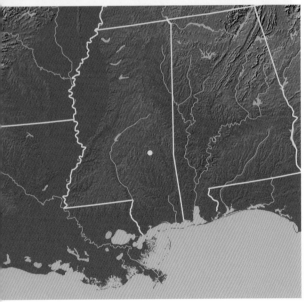

Bay Springs Salamander

Plethodon albagula Grobman, 1944

Western Slimy Salamander

Western Slimy Salamanders are distributed in two large areas: the uplands of the Ozark Plateau and the Ouachita Mountains of central and southern Missouri, northern Arkansas and extreme eastern Oklahoma, as well as on the Edwards Plateau and Balcones Escarpment in central Texas. They also occur in scattered localities in east Texas between these two large areas.

Western Slimy Salamanders are found in damp ravines, wooded hillsides, cave entrances, and wooded canyons, where they live under rocks, logs, or leaf litter..

Western Slimy Salamanders are listed as Protected by the State of Oklahoma.

ORIGINAL ACCOUNT by Carl D. Anthony; photograph by Dante Fenolio.

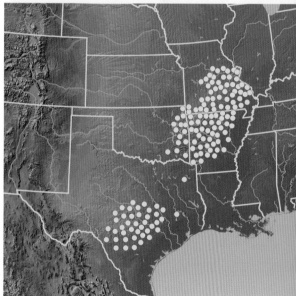

Western Slimy Salamander

Plethodon amplus Highton and Peabody, 2000

Blue Ridge Gray-cheeked Salamander

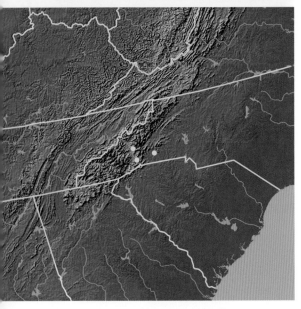

Blue Ridge Gray-cheeked Salamander

Blue Ridge Gray-cheeked Salamanders are known from the Blue Ridge Mountains of western North Carolina near Ashville at elevations of around 1110 m..

Blue Ridge Gray-cheeked Salamanders are reported to inhabit the forest floor and crevices in metamorphic rock.

Local extirpations of populations of Blue Ridge Gray-cheeked Salamanders have likely occurred because of habitat destruction and modification. Within their range, there are a few federal and state properties that contain suitable habitat for Blue Ridge Gray-cheeked Salamanders.

Blue Ridge Gray-cheeked Salamanders are not listed in North Carolina, the only state in which they occur.

ORIGINAL ACCOUNT by David A. Beamer and Michael J. Lannoo; photograph by Richard D. Bartlett.

Plethodon angusticlavius Grobman, 1944

Ozark Zigzag Salamander

Ozark Zigzag Salamanders are distributed throughout the Ozark Plateau from northwestern Arkansas to the eastern edge of Oklahoma into the southwestern portion of Missouri.

Ozark Zigzag Salamanders are found in moist leaf litter, talus areas, and caves. They prefer rocky habitats, and the availability of these rocky habitats therefore limits their abundance. Rocky cedar glades are used extensively by adult Ozark Zigzag Salamanders during the courtship season.

Little is known about the ecology of Ozark Zigzag Salamanders; therefore, risks to populations through perturbations of their moist woodland habitat can only be approximated.

Ozark Zigzag Salamanders are not listed either federally or at the state level in the U.S.

ORIGINAL ACCOUNT by Walter E. Meshaka Jr.; photograph by Michael Graziano.

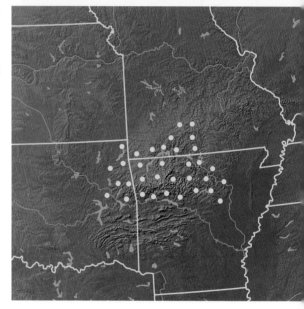

Ozark Zigzag Salamander

Plethodon asupak Mead, Clayton, Nauman, Olson, and Pfrender, 2005[15]

Scott Bar Salamander

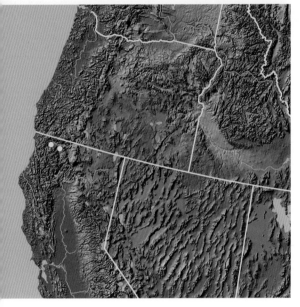

Scott Bar Salamander

Scott Bar Salamanders are distributed north of Scott Bar, near the confluence of the Scott and Klamath rivers in northern California, in an estimated range of 55,000 ha between Seiad Valley and Scott Bar Mountain.

Scott Bar Salamanders occur in forested habitats with rocky substrates. They are most predictably found near the ground surface during the spring and fall wet seasons when temperatures are above freezing. They retreat to subsurface burrows and cavities during dry and cold conditions.

The primary threats to Scott Bar Salamanders are disturbances that alter ground and subsurface refugia and moisture regimes. These may include intensive timber-harvesting that reduces canopy closure or compact substrates, mining activities, stand-replacement fires, and climate change.

Most of the range of Scott Bar Salamanders occurs on federal lands, where they are considered a Species of Interest by the U.S. Forest Service.

ACCOUNT by Deanna H. Olson; photograph by Gary Nafis.

Plethodon aureolus Highton, 1984

Tellico Salamander

Tellico Salamanders occur on the western slopes of the Unicoi Mountains and nearby lowlands in eastern Tennessee and western North Carolina, between the Little Tennessee and Hiwassee rivers. They have been found mostly at lower elevations but may go as high as 1622 m in North Carolina.

Tellico Salamanders occur in woodland detritus and retreat underground during cold or dry weather conditions. Juveniles have been found in abundance in late summer under superficial cover such as twigs.

Tellico Salamanders are relatively resilient to disturbances such as those associated with timber harvesting and are frequently found in second-growth forests. Within their range, there are many federal and state properties that contain suitable habitats for Tellico Salamanders.

Tellico Salamanders are not listed by either Tennessee or North Carolina, although Tennessee considers them to be a species "In Need of Management."

ORIGINAL ACCOUNT by David A. Beamer and Michael J. Lannoo; photograph by Richard D. Bartlett.

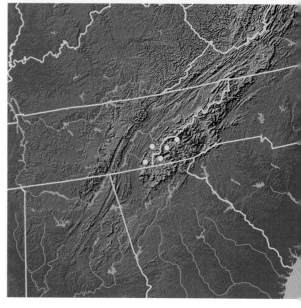

Tellico Salamander

Plethodon caddoensis Pope and Pope, 1951

Caddo Mountain Salamander

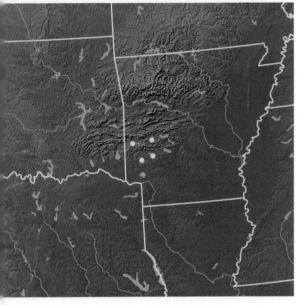

Caddo Mountain Salamander

Caddo Mountain Salamanders are locally distributed in the Caddo and Cossatot mountains of western Arkansas.

Caddo Mountain Salamanders are most commonly found under rocks, logs, and other forest debris at higher elevations of mixed deciduous, north-facing wooded slopes. Moisture conditions at the surface appear to greatly influence activity of Caddo Mountain Salamanders, and they retreat to lower levels of talus to escape hot and dry conditions. They also use caves and abandoned mines.

Caddo Mountain Salamanders are abundant within their limited range.

Caddo Mountain Salamanders are considered a Species of Special Concern in Arkansas.

ORIGINAL ACCOUNT by Carl D. Anthony; photograph by Mike Graziano.

Plethodon chattahoochee Highton, 1989

Chattahoochee Slimy Salamander

Chattahoochee Slimy Salamanders occur throughout much of the Chattahoochee National Forest in the Blue Ridge of northeastern Georgia and adjacent southwestern North Carolina.

Chattahoochee Slimy Salamanders, like other, similar species of salamanders, inhabit the forest floor and hide under assorted cover objects. They are likely to be most active under moist conditions and burrow to avoid inhospitably cold and dry surface conditions. Within their range, there are many federal and state properties that contain suitable habitats for Chattahoochee Slimy Salamanders.

Chattahoochee Slimy Salamanders are not listed in by either Georgia or North Carolina.

ORIGINAL ACCOUNT by David A. Beamer and Michael J. Lannoo; photograph by Gary Nafis.

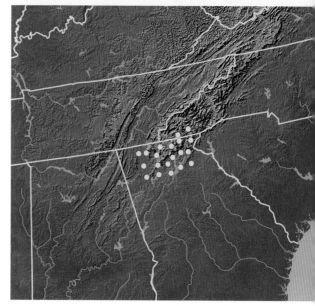

Chattahoochee Slimy Salamander

Plethodon cheoah Highton and Peabody, 2000

Cheoah Bald Salamander

Cheoah Bald Salamanders are known only from the Cheoah Bald in the Blue Ridge of western North Carolina at elevations of around 1400 m.

Cheoah Bald Salamanders inhabit moist deciduous forest. They likely move from the forest floor to underground sites with the onset of seasonally related cold or dry conditions, then back up to the forest floor with the return of favorable surface conditions.

Cheoah Bald Salamanders are not listed in North Carolina, the only state in which they occur.

ORIGINAL ACCOUNT by David A. Beamer and Michael J. Lannoo; photograph by Gary Nafis.

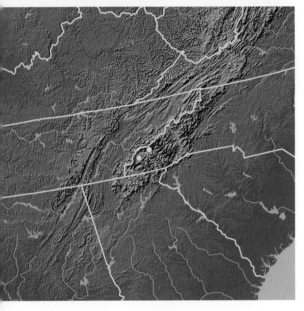

Cheoah Bald Salamander

Plethodon chlorobryonis Mittleman, 1951

Atlantic Coast Slimy Salamander

Atlantic Coast Slimy Salamanders occur in the Atlantic Coastal Plain of southeastern Virginia, eastern North Carolina, and northeastern South Carolina and up into the Piedmont of southeastern Virginia and central and western South Carolina and the Blue Ridge in northeastern Georgia.

Atlantic Coast Slimy Salamanders have been found in dry bottomlands along small creeks and under logs on slopes above cypress swamps. Adults are active nocturnally, especially under damp conditions. Atlantic Coast Slimy Salamanders likely avoid extremes of dry and cold by moving to underground sites.

There are many federal and state properties within their range that contain suitable habitats for Atlantic Coast Slimy Salamanders. The salamanders are relatively resilient to disturbances such as those associated with logging operations and they are frequently found in second-growth forests. Even in areas that have been converted to agricultural uses, populations of Atlantic Coast Slimy Salamanders still exist along larger, forested, stream corridors.

Atlantic Coast Slimy Salamanders are not listed by any state in the U.S.

ORIGINAL ACCOUNT by David A. Beamer and Michael J. Lannoo; photograph by Tim Herman.

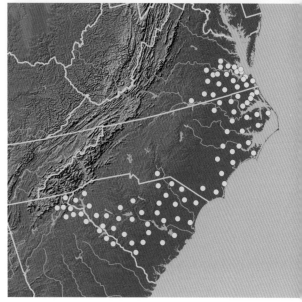

Atlantic Coast Slimy Salamander

Plethodon cinereus (Green, 1818)

Eastern Red-backed Salamander

Eastern Red-backed Salamander

Eastern Red-backed Salamanders range throughout the Maritime Provinces of Canada and southern Québec southward to western and northeastern North Carolina, northwestward to central Minnesota, and north of Lake Superior to Lake Nipigon in Ontario.

Eastern Red-backed Salamanders occupy deciduous, mixed coniferous–deciduous, and northern coniferous forests. They inhabit the leaf litter and retreat under stones, into soil cavities, and into rotting logs. Eastern Red-backed Salamanders have a limited ability to burrow, being effective only in soft substrates such as leaf litter or loose humus, and they prefer to use or enlarge existing retreats. They may also forage in bogs. Eastern Red-backed Salamanders prefer cool, moist microhabitats and avoid temperature extremes and desiccating environments. Most individuals reside below the soil surface. Eastern Red-backed Salamanders reach their highest abundance in mature hardwood forests with deep soils and abundant downed woody debris in various stages of decomposition. Juveniles often remain in the nest cavity with their mother for 1 to 3 weeks after hatching before they disperse.

Local extirpations of populations of Eastern Red-backed Salamanders due to habitat changes, chiefly deforestation, and other, unknown causes have been reported but, despite this, Eastern Red-backed Salamanders remain numerous.

Eastern Red-backed Salamanders are not listed under any provincial, state or federal regulations in either Canada or the U.S.

ORIGINAL ACCOUNT by Gary S. Casper; photograph by Mike Redmer.

Plethodon cylindraceus (Harlan, 1825)

White-spotted Slimy Salamander

White-spotted Slimy Salamanders occur in the Piedmont and Blue Ridge of Virginia and North Carolina, west to the French Broad River, and south to the Northern Piedmont of South Carolina. They also occur in extreme eastern West Virginia and in a small area of the Atlantic Coastal Plain of eastern Virginia.

White-spotted Slimy Salamanders occur in both virgin forest and second-growth oak–hickory forest with thick layers of leaf litter and abundant fallen logs. White-spotted Slimy Salamanders may be found beneath rotted logs but do not seem to occupy the leaf litter, even in wet periods.

White-spotted Slimy Salamanders have a wide distribution that includes many federal and state properties that contain suitable habitats. They are relatively resilient to disturbances such as those associated with timber harvesting and frequently are found in disturbed forests and relatively small, fragmented woodlots.

White-spotted Slimy Salamanders are not listed by any states within its distribution.

ORIGINAL ACCOUNT by David A. Beamer and Michael J. Lannoo; photograph by William Leonard.

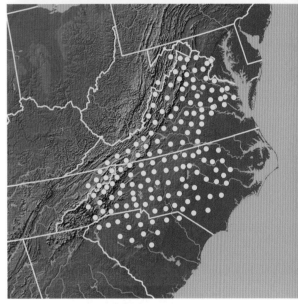

White-spotted Slimy Salamander

Plethodon dorsalis Cope, 1889

Northern Zigzag Salamander

Northern Zigzag Salamanders are distributed in the interior Low Plateau region from southwestern Indiana south through the Pennyrile region of Kentucky and the Cumberland Plateau of Tennessee to the Valley and Ridge region of central Alabama, as well as the hilly country of southwest Alabama west of the Alabama River.

Northern Zigzag Salamanders are found in areas of deciduous forest in moist leaf litter, talus and caves. They tend to prefer rocky habitats, which may limit their distribution and abundance.

Northern Zigzag Salamanders are listed as a Species of Special Concern in North Carolina, at the eastern edge of their distribution.

ORIGINAL ACCOUNT by Walter E. Meshaka Jr.; photograph by Mike Redmer.

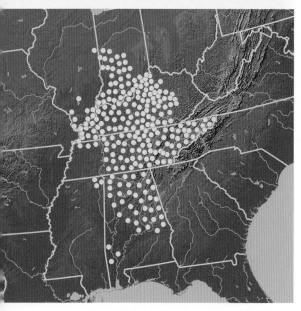

Northern Zigzag Salamander

Plethodon dunni Bishop, 1934

Dunn's Salamander

Dunn's Salamanders are distributed from the southwestern corner of Washington State southward through western Oregon, including the Oregon Cascades Range, and barely into extreme northwestern California. They are absent from the low parts of the Willamette Valley. Dunn's Salamanders occur from the high-tide line of the Pacific Ocean up the west flank of the Cascades to elevations of about 1,000 m.

Dunn's Salamanders are fairly widespread, but they appear to require wet rocky substrate and are most abundant in forest streamside habitats. They are almost always associated with rocks near seepages, springs, and streams. Along the Coast Range, they occur in sandstone or shale outcrops, whereas in the Cascades, they occur in basaltic talus. In rainy weather, they may be found in or under logs near streams or under surface debris.

There is no clear evidence suggesting population declines, but they are absent or reduced in abundance in clear-cut forests in western Oregon.

Dunn's Salamanders are a Species of Special Concern in California and are a candidate for state listing in Washington.

ORIGINAL ACCOUNT by R. Bruce Bury; photograph by William Leonard.

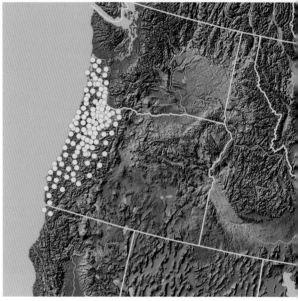

Dunn's Salamander

Plethodon electromorphus Highton, 1999

Northern Ravine Salamander

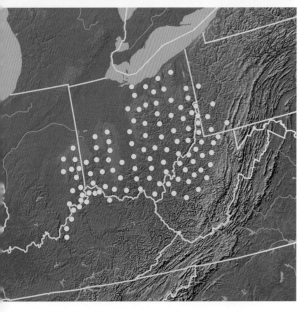

Northern Ravine Salamanders are found in southwestern Pennsylvania through most of Ohio to southeastern Indiana and south to northwestern West Virginia and northern Kentucky.

Northern Ravine Salamanders have been found on forested slopes with friable soil and flat rocks. They show a strong preference for rocks and are rarely found under any other type of cover. Talus slopes sometimes harbor large numbers of individuals. Northern Ravine Salamanders avoid both the dry crests of ridges and excessively moist situations such as stream margins. Hatchlings remain underground, possibly at the nest site, for several months. Within their range, there are many federal and state properties that contain suitable habitats for these salamanders.

Northern Ravine Salamanders are not listed by any state in their distribution.

ORIGINAL ACCOUNT by David A. Beamer and Michael J. Lannoo; photograph by Michael Graziano.

Northern Ravine Salamander

Plethodon elongatus Van Denburgh, 1916

Del Norte Salamander

Del Norte Salamanders are found on the west side of the Coastal Range in northwestern California and southwestern Oregon from approximately Cape Blanco south near to Arcata Bay and inland to the Siskiyou Mountains. They also have been found along West Cow Creek in the Umpqua River watershed in Oregon.

Adult Del Norte Salamanders are mostly associated with talus or rocky substrates, as well as with downed woody debris in areas with nearby rock substrates. They prefer older-aged forest stands with enclosed canopies, though in damp coastal areas this is less of a requirement.

At one time, Del Norte Salamanders were a Survey and Manage Species that was afforded protection on federal lands under the Northwest Forest Plan, where ground-disturbing activities were restricted on occupied sites and a surrounding buffer. However, Del Norte Salamanders are not currently listed at either the state or federal level in the U.S.

ORIGINAL ACCOUNT by Hartwell H. Welsh Jr. and R. Bruce Bury; photograph by William Leonard.

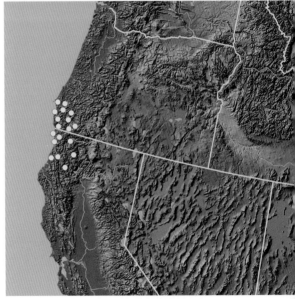

Del Norte Salamander

Plethodon fourchensis Duncan and Highton, 1979

Fourche Mountain Salamander

Fourche Mountain Salamanders are found only on Fourche, eastern Iron Fork, and Shut-In Mountains in the Ouachitas of west central Arkansas just north of the city of Mena.

Fourche Mountain Salamanders are most commonly found at higher elevations in mixed deciduous forest on north-facing wooded slopes, especially in deep ravines. They are most abundant where there is high plant diversity, forest canopy cover, and overall wetness. Fourche Mountain Salamanders use rocks, logs, and other forest debris as cover objects.

Fourche Mountain Salamanders have a limited distribution, yet within this range they can be abundant, although there has been a reduction in numbers at several localities.

Fourche Mountain Salamanders are not listed either federally or at the state level in the U.S.

ORIGINAL ACCOUNT by Carl D. Anthony; photograph by Michael Graziano.

Fourche Mountain Salamander

Plethodon glutinosus (Green, 1818)

Northern Slimy Salamander

Northern Slimy Salamanders are present from northeastern Alabama, northern Georgia, and extreme southwestern North Carolina northeast along the Appalachian Mountains into southwestern Connecticut and southern New York and west into eastern Ohio, southern and western Indiana, and southern Illinois. There is also a disjunct population in southern New Hampshire. Northern Slimy Salamanders occur at elevations from sea level to about 1500 m.

Northern Slimy Salamanders are to be found beneath logs, stones, moist humus and leaf mold, or in manure piles in woods, in the crevices of shale banks, along the sides of gullies and ravines, and even in caves in Alabama and Illinois. They reach their greatest abundance in mature hardwood forests.

Northern Slimy Salamanders are relatively resilient to disturbances caused by logging operations and are frequently found in second-growth forests and relatively small isolated woodlots. Many federal and state properties contain suitable habitats for Northern Slimy Salamanders.

Northern Slimy Salamanders are listed as Threatened in Connecticut and as Protected in New Jersey.

ORIGINAL ACCOUNT by David A. Beamer and Michael J. Lannoo; photograph by Mike Redmer.

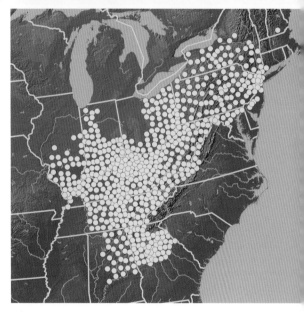

Northern Slimy Salamander

Plethodon grobmani Allen and Neill, 1949

Southeastern Slimy Salamander

Southeastern Slimy Salamanders are distributed from southern Alabama and southern Georgia south to central Florida.

Southeastern Slimy Salamanders are common in steephead ravines, maritime forests, and river-bottom hardwood forests. Much of the area within the range of Southeastern Slimy Salamanders historically was covered with savanna and prairie, and in these areas they are rare or absent. Southeastern Slimy Salamanders do not require pristine habitats or old-growth forests and are often found under discarded rubbish, though they are absent from small woodlots. Juvenile Southeastern Slimy Salamanders likely stay at the nest site for the first 2 months after hatching.

Southeastern Slimy Salamanders are reasonably resistant to the habitat disruptions associated with forestry activities and may frequently be found in second-growth forests.

Southeastern Slimy Salamanders are not listed in any of the states in which they occur.

ORIGINAL ACCOUNT by David A. Beamer and Michael J. Lannoo; photograph by Aubrey Huepel.

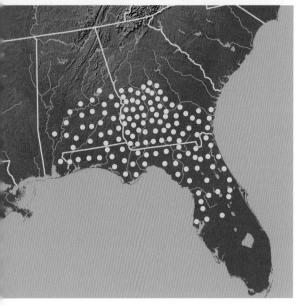

Southeastern Slimy Salamander

Plethodon hoffmani Highton, 1971

Valley and Ridge Salamander

Valley and Ridge Salamanders occur in the Appalachian Plateau and Allegheny Mountains of central Pennsylvania and south through the intervening Valley and Ridge province to the Blue Ridge of Virginia and West Virginia as far as the New River.

Valley and Ridge Salamanders inhabit hillside slopes of mixed deciduous forest with flat stones. Most of the terrain within their range is dry and well drained. Valley and Ridge Salamanders will enter caves, where they reside beneath cover in the twilight zone. Juveniles likely stay at the nest site for several months after hatching and appear on the surface in March.

Some populations of Valley and Ridge Salamanders in central Pennsylvania and northern Virginia may have declined drastically.

Valley and Ridge Salamanders are not listed by any state in the U.S.

ORIGINAL ACCOUNT by David A. Beamer and Michael J. Lannoo; photograph by John Clare.

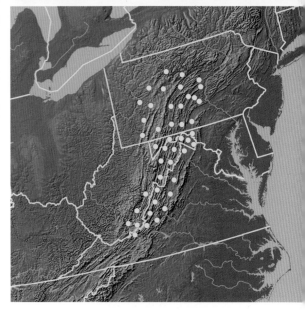

Valley and Ridge Salamander

Plethodon hubrichti Thurow, 1957

Peaks of Otter Salamander

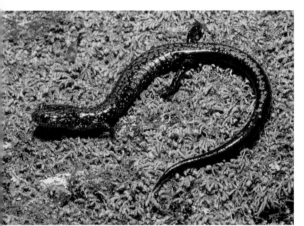

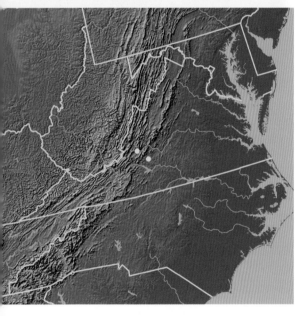

Peaks of Otter Salamander

Peaks of Otter Salamanders are restricted to the Peaks of Otter, an approximately 16-km length of the Blue Ridge Mountains in central Virginia.

Peaks of Otter Salamanders occupy the forest floor of mature Appalachian hardwood forest at elevations above 550 m. They have been found among a variety of vegetation types, including rhododendron thickets. The highest densities of Peaks of Otter Salamanders occur in areas of deep, moist soil in mature hardwood forest stands with abundant rocks and logs.

The U.S. Forest Service and the U.S. National Park Service specify the entire range of the Peaks of Otter Salamander as a Special Biological Area with a primary conservation area that allows no logging activities and a secondary conservation area that allows logging to take place under certain restrictions.

Peaks of Otter Salamanders are recognized as a federal species At Risk by the U.S. Fish and Wildlife Service and are listed as a Species of Concern by the Virginia Department of Game and Inland Fisheries. The U.S. Forest Service lists them as a Sensitive Species.

ORIGINAL ACCOUNT by Joseph C. Mitchell and Jill A. Wicknick; photograph by Gary Nafis.

Plethodon idahoensis Slater and Slipp, 1940

Coeur d'Alene Salamander

Coeur d'Alene Salamanders are locally distributed at elevations between 500 and 1550 m in the Bitterroot Range and Clearwater Mountains of northern Idaho and northwestern Montana and north into the Purcell and Selkirk mountains of southeastern British Columbia as far as Mica Creek, 95 km north of the town of Revelstoke.

Coeur d'Alene Salamanders are most commonly found in semiaquatic habitats, including seepages, streamside talus, and splash zones, but may also occur in forest debris and amid damp talus.

Coeur d'Alene Salamanders are considered a state species of Special Concern in both Idaho and Montana. They are a species of Special Concern in Canada under the federal Species at Risk Act and are protected under British Columbia's Wildlife Act.

ORIGINAL ACCOUNT by Kirk Lohman; photograph by William Leonard.

Coeur d'Alene Salamander

Plethodon jordani Blatchley, 1901

Jordan's Salamander

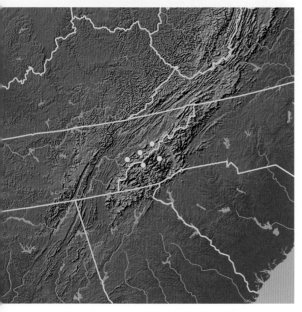

Red-cheeked Salamander

Jordan's Salamanders are found in the Great Smoky Mountains that straddle the Tennessee–North Carolina border. They occur from the highest peak in the Great Smoky Mountains at 2024 m down to an elevation of 213 m.

Jordan's Salamanders are most abundant in forests dominated by red spruce and Fraser's fir but are also found on hardwood-covered ridges. Jordan's Salamanders are most abundant where the forest floor is covered with a heavy layer of moss, with only a little soil over a mass of large boulders. Jordan's Salamanders inhabit burrows or other subterranean passageways under rocks, logs, and other cover objects during warmer months. These burrow systems can be extensive. Jordan's Salamanders are most active at night and during and following rains. Juveniles remain underground after hatching for 10 to 12 months until the following summer. They can be found beneath rocks in the midst of a thick carpet of fallen needles and occasionally beneath pieces of wood, including rotten logs and bark.

Jordan's salamanders are not listed by either of the states in which they occur but they are protected within Great Smoky Mountains National Park.

ORIGINAL ACCOUNT by David A. Beamer and Michael J. Lannoo; photograph by Mike Redmer.

Plethodon kentucki Mittleman, 1951

Cumberland Plateau Salamander

Cumberland Plateau Salamanders are found in the Cumberland Plateau of eastern Kentucky, northeastern Tennessee, southwestern Virginia, and West Virginia west of the New and Kanawha Rivers.

Cumberland Plateau Salamanders are found in mature hardwood forests and occupy a variety of woodland habitats, such as moist ravines and hillsides, shale banks, cave entrances, and rock crevices. They use rocks preferentially as cover objects and are found most frequently on west-facing slopes.

Forestry practices that include harvesting of hardwoods and aggressive rotation schedules likely have negative impacts on Cumberland Plateau Salamanders.

Cumberland Plateau Salamanders lack listed status at both state and federal levels.

ORIGINAL ACCOUNT by Thomas K. Pauley and Mark B. Watson; photograph by Mike Redmer.

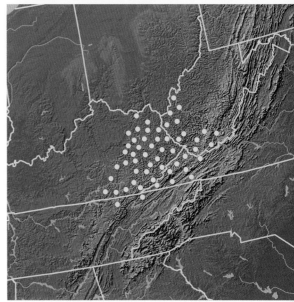

Cumberland Plateau Salamander

Plethodon kiamichi Highton, 1989

Kiamichi Slimy Salamander

Kiamichi Slimy Salamanders are known only from Kiamichi and Round Mountains in the Ouchita Mountains of southeastern Oklahoma and western Arkansas.

Kiamichi Slimy Salamanders are most commonly found at higher elevations in moist woods and ravines. They are typically found under rocks and logs.

Though they have a limited distribution, Kiamichi Slimy Salamanders can be locally abundant.

Kiamichi Slimy Salamanders are listed as Protected by the state of Oklahoma.

ORIGINAL ACCOUNT by Carl D. Anthony; photograph by Michael Graziano.

Kiamichi Slimy Salamander

Plethodon kisatchie Highton, 1989

Louisiana Slimy Salamander

Louisiana Slimy Salamanders are known from the hill parishes of north central Louisiana north into southern Arkansas.

Louisiana Slimy Salamanders are most often found in hardwood forests but will also occur in forests where pines predominate. In these habitats, Louisiana Slimy Salamanders are well established but not common.

Louisiana Slimy Salamanders are not listed at either the state or federal level in the U.S.

ORIGINAL ACCOUNT by Carl D. Anthony; photograph by Richard D. Bartlett.

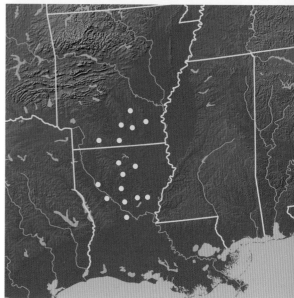

Louisiana Slimy Salamander

Plethodon larselli Burns, 1954

Larch Mountain Salamander

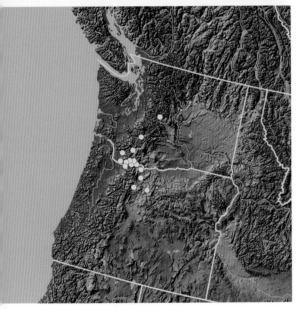

Larch Mountain Salamanders are limited to a narrow region in southwest Washington and northwest Oregon where the Cascades Range has been eroded by the Columbia River in the Columbia River Gorge.

Larch Mountain Salamanders typically are associated with steep, at least partially forested, talus slopes, though they are also occasionally found in other habitats, such as entrances to ice caves and within forests no longer associated with exposed talus. In optimal habitat, Larch Mountain Salamanders remain relatively common.

Larch Mountain Salamanders are considered by Washington and Oregon to be a Sensitive species.

ORIGINAL ACCOUNT by Robert E. Herrington; photograph by Brad Moon.

Larch Mountain Salamander

Plethodon meridianus Highton and Peabody, 2000

South Mountain Gray-cheeked Salamander

South Mountain Gray-cheeked Salamanders are found only on and around South Mountain in the Piedmont of central North Carolina.

South Mountain Gray-cheeked Salamanders preferentially inhabit mature closed-canopy forests, where they seek shelter under logs, rocks, or other cover objects during the day and are active on the forest floor at night. They are more active under moist conditions and likely avoid dry and cold extremes by moving to underground sites.

Although some habitats used South Mountain Gray-cheeked Salamanders are preserved as game lands or lie within South Mountain State Park, much of the land has been converted to housing, and it is likely that more of their habitat will continue to be lost in the near future.

South Mountain Gray-cheeked Salamanders are not listed by North Carolina, the only state in which they occur.

ORIGINAL ACCOUNT by David A. Beamer and Michael J. Lannoo; photograph by Steve Tilley.

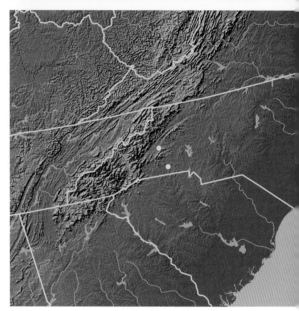

South Mountain Gray-cheeked Salamander

Plethodon metcalfi Highton and Peabody, 2000

Southern Gray-cheeked Salamander

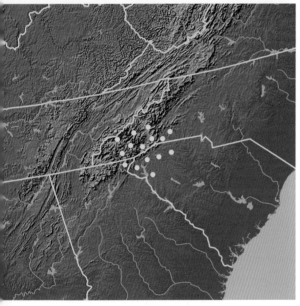

Southern Gray-cheeked Salamander

Southern Gray-cheeked Salamanders are found in the southern Blue Ridge Mountains of North Carolina, South Carolina, and northeastern Georgia. Though one population occurs at an elevation of 256 m, most populations of Southern Gray-cheeked Salamander are at elevations upward from 750 m to the limits of large-tree growth.

Southern Gray-cheeked Salamanders are found in oak–chestnut forest in areas with good, deep soil. They hide beneath logs or wet leaf litter. At night, Southern Gray-cheeked Salamanders may be on found on rock faces near the intersection of stone and leaf litter and, on rainy nights, on the leaves of herbaceous vegetation up to 30 cm above the ground.

Southern Gray-cheeked Salamanders are relatively resilient to disturbances, such as those associated with timber harvesting, and frequently are found in second-growth forests and relatively small, isolated woodlots. However, they have been known to disappear from clear-cut areas within 4 years of logging. A portion of their original distribution in South Carolina is now underwater, covered by Lake Jocassee, and therefore uninhabitable.

Southern Gray-cheeked Salamanders are not listed by any state in which they occur.

ORIGINAL ACCOUNT by David A. Beamer and Michael J. Lannoo; photograph by Mike Redmer.

Plethodon mississippi Highton, 1989

Mississippi Slimy Salamander

Mississippi Slimy Salamanders are distributed east of the Mississippi River from southwestern Kentucky south through western Tennessee, western Alabama, and most of Mississippi to southeastern Louisiana.

Mississippi Slimy Salamanders are common in maritime forest and river-bottom hardwood forests, including oak–hickory forest. They can be found in second-growth forests but tend to be absent from savanna and prairie habitats and from small woodlots. They can found under assorted types of cover objects within the forest and even under discarded rubbish. Juveniles likely stay at the nest site for the first several months after hatching.

Mississippi Slimy Salamanders do not require pristine habitats or old-growth forests. There are several federal and state properties that contain suitable habitats for them.

Mississippi Slimy Salamanders are not listed by any state.

ORIGINAL ACCOUNT by David A. Beamer and Michael J. Lannoo; photograph by Brad Moon.

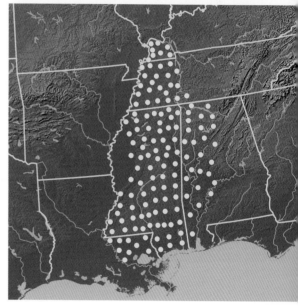

Mississippi Slimy Salamander

Plethodon montanus Highton and Peabody, 2000

Northern Gray-cheeked Salamander

Northern Gray-cheeked Salamanders occur at higher elevations on mountains in the Blue Ridge of western North Carolina, eastern Tennessee, and southwest Virginia, as well several as mountaintops in the Ridge and Valley region of western Virginia.

Northern Gray-cheeked Salamanders have been recorded from both virgin and second-growth mixed forest with canopies that may include red and white oak, maples, buckeyes, ironwood, gum, tulip trees, and hemlocks. Northern Gray-cheeked Salamanders are also found in high-elevation spruce–fir forest, where they are found under logs but not in stumps.

As Northern Gray-cheeked Salamanders are restricted to higher elevations, suitable habitat for them may be separated by stretches of lower uninhabitable areas, and populations often are not continuous.

Northern Gray-cheeked Salamanders are not listed by any state.

ORIGINAL ACCOUNT by David A. Beamer and Michael J. Lannoo; photograph by Nathan Haislip.

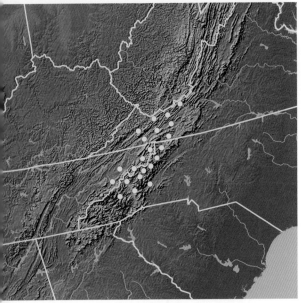

Northern Gray-cheeked Salamander

Plethodon neomexicanus Stebbins and Riemer, 1950

Jemez Mountains Salamander

Jemez Mountains Salamanders are known only from high elevations—generally between 2200 and 2900 m, though they may occur as high as 3429 m—in the mountains surrounding the caldera of the Valle Grande of the Jemez Mountains in north central New Mexico. Their range is limited to an area of approximately 971 km².

Jemez Mountains Salamanders occur in coniferous forest dominated by Douglas fir, blue spruce, Engelmann spruce, ponderosa pine, and white fir, with occasional aspen. Deciduous understory shrubs include Rocky Mountain maple, New Mexico locust, oceanspray, and various shrubby oaks. Jemez Mountains Salamanders are mostly found on loose rocky soils in and under rotting Douglas fir logs, or under rocks on flat areas and steep slopes.

Populations of Jemez Mountains Salamanders are highly localized because of the distribution of suitable subsurface geology. Though, overall, Jemez Mountains Salamanders are rare, they may be locally numerous in small areas of suitable habitat.

Jemez Mountains Salamanders are listed as Threatened by the State of New Mexico and as a Species of Concern by the U.S. Forest Service and the U.S. Fish and Wildlife Service.

ORIGINAL ACCOUNT by Charles W. Painter; photograph by Gary Nafis.

Jemez Mountains Salamander

Plethodon nettingi Green, 1938

Cheat Mountain Salamander

Cheat Mountain Salamanders are limited to the eastern highlands of West Virginia on and near Cheat Mountain at elevations above 750 m.

Cheat Mountain Salamanders are found in spruce, hemlock, or deciduous forest stands with scattered spruce and hemlock. They also occur in boulder fields, rock outcrops, and steep ravines lined with dense growth of rhododendron.

Cheat Mountain Salamanders may be sensitive to logging practices such as clear-cutting and the development of roads, trails, and ski slopes. However, most of the known populations of Cheat Mountain Salamanders are located within the boundaries of the Monongahela National Forest, which provides protection from habitat disturbances.

Cheat Mountain Salamanders are listed as a Threatened species by the U.S. Fish and Wildlife Service.

ORIGINAL ACCOUNT by Thomas K. Pauley, Beth Anne Pauley, and Mark B. Watson; photograph by Michael Graziano.

Cheat Mountain Salamander

Plethodon ocmulgee Highton, 1989

Ocmulgee Slimy Salamander

Ocmulgee Slimy Salamanders are found in the upper Coastal Plain and adjacent Piedmont district of east central Georgia associated with the Ocmulgee River drainage.

Ocmulgee Slimy Salamanders inhabit the forest floor, where during the daytime, they will use logs and rocks as cover objects or move into underground retreats. Ocmulgee Slimy Salamanders tend to be nocturnal, and their activity levels likely are related to the availability of sufficient moisture.

There are few federal and state properties that preserve suitable habitat for Ocmulgee Slimy Salamanders.

Ocmulgee Slimy Salamanders are not listed by Georgia, the only state in which they occur.

ORIGINAL ACCOUNT by David A. Beamer and Michael J. Lannoo; photograph by William Leonard.

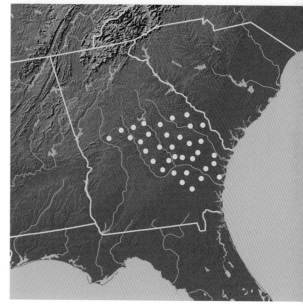

Ocmulgee Slimy Salamander

Plethodon ouachitae Dunn and Heinze, 1933

Rich Mountain Salamander

Rich Mountain Salamanders are distributed in the Ouachita Mountains of eastern Oklahoma and western Arkansas, including parts of Winding Stair, Rich, and Kiamichi mountains, as well as other nearby peaks.

Rich Mountain Salamanders are most commonly found at higher elevations in mixed deciduous woods on north-, northeast- and northwest-facing slopes, often adjacent to seeps. Sandstone rocks, logs, and other forest debris are commonly used by Rich Mountain Salamanders as cover objects. Moisture conditions at the surface greatly influence the activity of Rich Mountain Salamanders, and they will retreat to lower levels of talus to escape hot and dry conditions. They may also be found in cave entrances.

Rich Mountain Salamanders are listed as a Protected species by the state of Oklahoma.

ORIGINAL ACCOUNT by Carl D. Anthony; photograph by Dante Fenolio.

Rich Mountain Salamander

Plethodon petraeus Wynn, Highton, and Jacobs, 1988

Pigeon Mountain Salamander

Pigeon Mountain Salamanders are known only from the eastern slope of Pigeon Mountain in northwestern Georgia.

Pigeon Mountain Salamanders are associated with limestone outcroppings, boulder fields, and caves within moist oak–hickory forests. They are most often found in and around cracks and crevices within rocks. Those found in caves are rarely deeper than the twilight zone.

The restricted distribution of Pigeon Mountain Salamanders makes them especially vulnerable to threats to their habitats. However, the vast majority of their range is under public ownership as the Crockford-Pigeon Mountain Wildlife Management Area.

Pigeon Mountain Salamanders are listed as Rare and thus are protected by the State of Georgia.

ORIGINAL ACCOUNT by John B. Jensen and Carlos D. Camp; photograph by Tim Herman.

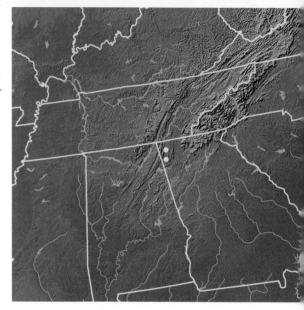

Pigeon Mountain Salamander

Plethodon punctatus Highton, "1971," 1972

Cow Knob Salamander

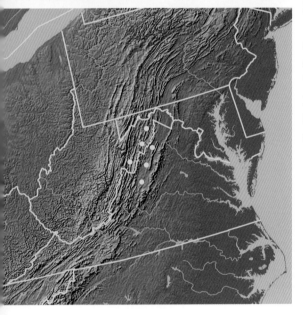

Cow Knob Salamander

Cow Knob Salamanders are restricted to elevations above about 730 m in the Shenandoah Mountains and Blue Ridge Mountains of northwestern Virginia and eastern West Virginia.

Cow Knob Salamanders occur in hemlock stands, old-growth hardwood forests, and mature hardwood forests. They are most commonly found under rocks in moist areas in deep soil on north-facing slopes above elevations of 900 m. Most of the land above 900 m within the range of Cow Knob Salamanders has been designated as the Shenandoah Mountain Crest Special Biological Area and is off limits to logging.

Cow Knob Salamanders are listed as a Species at Risk by the U.S. Fish and Wildlife Service and a Species of Special Concern in Virginia and West Virginia. The George Washington National Forest, in which most of the range of Cow Knob Salamanders occurs, recognizes them as a Sensitive species.

ORIGINAL ACCOUNT by Joseph C. Mitchell and Thomas K. Pauley; photograph by Michael Graziano.

Plethodon richmondi Netting and Mittleman, 1938

Southern Ravine Salamander

The range of Southern Ravine Salamanders lies in the Blue Ridge Mountains, Appalachian Valley and Ridge region, and Cumberland Plateau of eastern Kentucky, western Virginia and southern West Virginia, extreme northeastern Tennessee, and extreme northwestern North Carolina. The distribution in the western portion of their range is strictly south of the Kanawha and Ohio rivers. Southern Ravine Salamanders reach elevations of 1300 m on Big Black Mountain in Kentucky.

Southern Ravine Salamanders prefer steep to sloping hillsides and ravines covered by damp forests, with flat rocks, rock outcrops, logs, and abundant leaf litter under which the salamanders hide. Southern Ravine Salamanders are less common in floodplains or on dry ridge tops but are found occasionally in pastures adjacent to wooded areas.

Southern Ravine Salamanders are listed as a species of Scientific Interest in Tennessee because of their limited distribution.

ORIGINAL ACCOUNT by Thomas K. Pauley and Mark B. Watson; photograph by William Leonard.

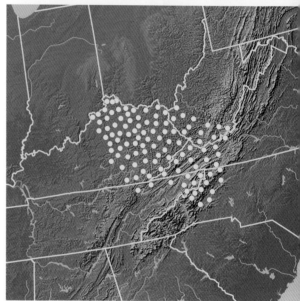

Southern Ravine Salamander

Plethodon savannah Highton, 1989

Savannah Slimy Salamander

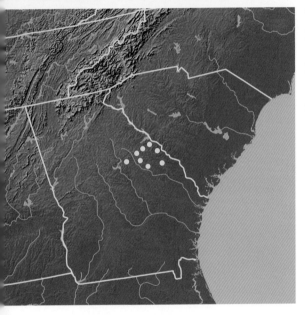

Savannah Slimy Salamander

Savannah Slimy Salamanders are known only from low hills between Brier Creek and the Savannah River in east central Georgia, east of the city of Waynesboro.

Savannah Slimy Salamanders inhabit the floor of deciduous forest and hide under logs and rocks during daylight hours. They are active at night, especially under moist conditions. Savannah Slimy Salamanders likely avoid conditions of unsuitable cold and dryness by seeking shelter in underground sites.

Within their small range, there do not appear to be any federal or state properties that would preserve suitable habitats for Savannah Slimy Salamanders, which leaves them vulnerable to habitat loss.

Savannah Slimy Salamanders are not listed in Georgia.

ORIGINAL ACCOUNT by David A. Beamer and Michael J. Lannoo; photograph by Pierson Hill.

Plethodon sequoyah Highton, 1989

Sequoyah Slimy Salamander

The only known localities for Sequoyah Slimy Salamanders are Beavers Bend State Park in southeastern Oklahoma and De Queen Lake in nearby western Arkansas.

Sequoyah Slimy Salamanders occur in moist woods and ravines, under various rocks and woody debris. They are locally numerous within their extremely limited range.

Sequoyah Slimy Salamanders are not specifically listed by the states of Arkansas or Oklahoma.

ORIGINAL ACCOUNT by Carl D. Anthony; photograph by Michael Graziano.

Sequoyah Slimy Salamander

Plethodon serratus Grobman, 1944

Southern Red-backed Salamander

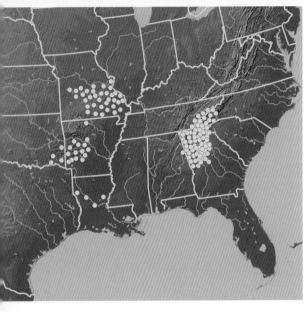

Southern Red-backed Salamander

Southern red-backed Salamanders occur in four distinct and isolated areas: the Ozark Plateau of southeastern Missouri; the southern Blue Ridge, Great Smoky Mountains and Piedmont of northwestern Georgia, eastern Alabama, eastern Tennessee, and western North Carolina, the Ouachita Mountains of southeastern Oklahoma and western Arkansas, and central Louisiana in the valley of the Red River. Populations occur at elevations up to 1690 m.

Southern Red-backed Salamanders are found in a variety of habitats. In Louisiana, they occur on slopes littered with sandstone rocks and with sandy/clay soils whereas, in the Great Smoky Mountains, they are found in pine–oak or mixed hardwood forests and, in the Georgia Piedmont, they inhabit hilly terrain with frequent steep bluffs covered by moist deciduous forest. In each of these habitats, Southern Red-backed Salamanders favor slopes having fairly dry soil and are found in leaf duff, under logs, in leaf packs along the margins of small streams or seeps, beneath rocks, and in leaf litter and stump holes.

Federal and state properties containing suitable habitat for Southern Red-backed Salamanders occur in most of the various parts of their range, though populations living in areas where hardwood forests have been converted to intensively managed pine forests have suffered.

Southern Red-backed Salamanders are listed as a Species of Special Concern in Louisiana.

ORIGINAL ACCOUNT by David A. Beamer and Michael J. Lannoo; photograph by Dante Fenolio.

Plethodon shenandoah Highton and Worthington, 1967

Shenandoah Salamander

Shenandoah Salamanders are known only from Hawksbill Mountain, Stony Man Mountain and The Pinnacles in the Blue Ridge of north central Virginia at elevations above 900 m.

Shenandoah Salamanders occupy relatively dry, north-facing talus slopes that support varying densities of hardwood trees. The salamanders live among the moist soil pockets and rocks that characterize these habitats.

Shenandoah Salamanders are threatened by alteration of forest canopy cover due to defoliation by the introduced gypsy moth, by acid precipitation, and by overgrowth of their talus habitat. They suffer increased risk of mortality if environmental perturbations cause drying in the talus.

Shenandoah Salamanders have been listed as Endangered by the Virginia Department of Game and Inland Fisheries and by the U.S. Fish and Wildlife Service.

ORIGINAL ACCOUNT by Joseph C. Mitchell; photograph by Gary Nafis.

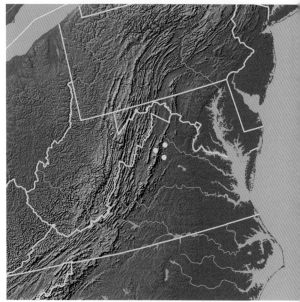

Shenandoah Salamander

Plethodon sherando Highton, 2004

Big Levels Salamander

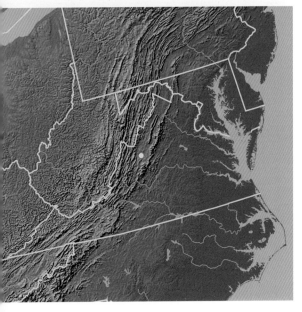

Big Levels Salamander

Big Levels Salamanders occur in the Big Levels area of the Blue Ridge Mountains in north central Virginia at elevations between 579 m at Lake Sherando to 1091 m on the peak of Bald Mountain.

Big Levels Salamanders are found in rocky talus slopes and in hardwood forest, where they hide under dead leaves and bark.

Most of the range of Big Levels Salamanders is within the George Washington National Forest. The apparently small distribution of Big Levels Salamanders is inherently a conservation concern but they are not listed by the state of Virginia.

ORIGINAL ACCOUNT by Michael J. Lannoo; photograph by John Clare.

Plethodon shermani Brimley, 1912

Red-legged Salamander

Red-legged Salamanders are found in the Unicoi and Nantahala mountains of North Carolina and southeastern Tennessee, as well as on and around Rabun Bald in extreme northeastern Georgia.

Red-legged Salamanders inhabit maple and birch forest and may be found beneath stones, inside and under rotten logs, under sticks and loose bark, and beneath solid logs in places such as rhododendron thickets bordering little streams.

Red-legged Salamanders appear to be are relatively resistant to disturbances such as those associated with logging operations. Some populations remain robust, but in others, Red-legged Salamanders have dropped in abundance.

Red-legged Salamanders are not specifically listed in North Carolina or Georgia.

ORIGINAL ACCOUNT by David A. Beamer and Michael J. Lannoo; photograph by Dante Fenolio.

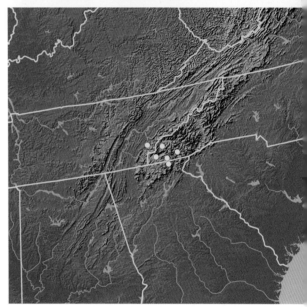

Red-legged Salamander

Plethodon stormi Highton and Brame, 1965

Siskiyou Mountains Salamander

Siskiyou Mountains Salamander

Siskiyou Mountains Salamanders are restricted to the Siskiyou Mountains in the upper Klamath River watershed of northern California, and the adjacent Applegate River watershed in southwest Oregon.

The largest populations of Siskiyou Mountains Salamanders are found in heavily wooded, north-facing slopes with rocky talus. Siskiyou Mountains Salamanders of all sizes and sexes occur in talus and rocky soils and occasionally are found under logs, leaf litter, and other cover if talus is nearby.

There has been considerable loss and fragmentation of the habitat of Siskiyou Mountains Salamanders due to past logging practices.

Once considered a U.S. federal Candidate Species for listing, Siskiyou Mountains Salamanders are recognized as a species of Special Concern by both Oregon and California and are listed as a Survey and Manage Species under the Northwest Forest Plan on federal lands. This stipulates that ground-disturbing activities are not permitted where Siskiyou Mountains Salamanders occur or within a 33 m buffer around suitable habitat.

ORIGINAL ACCOUNT by R. Bruce Bury and Hartwell H. Welsh Jr.; photograph by William Leonard.

Plethodon teyahalee Hairston, 1950

Southern Appalachian Salamander

Southern Appalachian Salamanders are distributed in the southern Blue Ridge Mountains of southwestern North Carolina and immediately adjacent Tennessee, South Carolina, and Georgia at elevations up to 1550 m.

Southern Appalachian Salamanders typically are found under rocks and logs in deciduous forests and emerge at night to forage and feed on the forest floor. Juveniles are found in the same area as adults, but frequently occur under smaller superficial cover objects, such as twigs and bits of bark. Southern Appalachian Salamanders are most active when it is damp.

Evidence of declines in abundance has been found in some populations of Southern Appalachian Salamanders. This may reflect either true declines or natural population fluctuations.

Southern Appalachian Salamanders are not listed by any of the states in which they occur.

ORIGINAL ACCOUNT by David A. Beamer and Michael J. Lannoo; photograph by Gary Nafis.

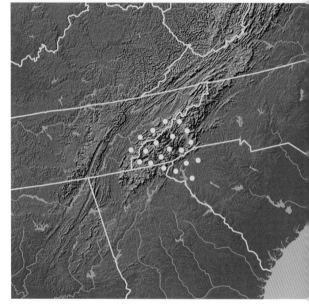

Southern Appalachian Salamander

Plethodon vandykei Van Denburgh, 1906

Van Dyke's Salamander

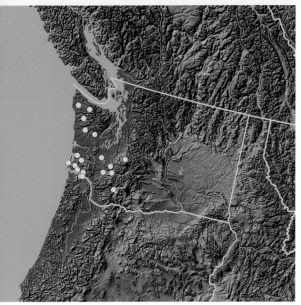

Van Dyke's Salamander

Van Dyke's Salamanders are found in several isolated localities at elevations from sea level to 1560 m in western and west central Washington, including the Olympic Mountains, the Willapa Hills, the Cascades Range, and the Chehalis Valley south of Puget Sound.

Van Dyke's Salamanders generally are found in coniferous forests in regions that receive over 1.5 m of annual rainfall and have an upper limit of elevation at the lower edge of subalpine forests. In coastal habitats, Van Dyke's Salamanders are found associated with rocks or woody debris, whereas in interior forests they generally are found associated with moist talus on north-facing slopes. Van Dyke's Salamanders are found beneath moist stones and moss near running water and seeps. Adults are most active on the surface of the forest floor in spring and fall.

Though many populations of Van Dyke's Salamanders are on National Park Service lands, and are therefore afforded some degree of protection, they do not fare well in intensively managed forests.

Van Dyke's Salamanders are rare and protected as a State Candidate species for listing by the State of Washington. Federally, the U.S. lists them as a Species of Concern.

ORIGINAL ACCOUNT by David A. Beamer and Michael J. Lannoo; photograph by Brad Moon.

Plethodon variolatus (Gilliams, 1818)

South Carolina Slimy Salamander

South Carolina Slimy Salamanders occur in the southern Atlantic Coastal Plain of South Carolina and eastern Georgia.

South Carolina Slimy Salamanders may be found in mixed hardwood forests, bottomland forests, and longleaf pine savannas but appear to avoid low sandy areas with palmetto and pine flatwoods. South Carolina Slimy Salamanders prefer thick, damp woods where the ground is littered with bark scraps, fallen timber, and leafy debris. They may also use shrew burrows as retreats. During early spring, juvenile South Carolina Slimy Salamanders can be encountered in piles of pine bark surrounding naked pine boles.

South Carolina Slimy Salamanders are frequently found in areas that had undergone prescribed burning as a land-management practice. Unburned areas are always present nearby if South Carolina Slimy Salamanders occur in these areas, so it is possible that the salamanders find refuge in the unburned sites.

South Carolina Slimy Salamanders are not listed by either of the two states in which they occur.

ORIGINAL ACCOUNT by David A. Beamer and Michael J. Lannoo; photograph by Todd Pierson.

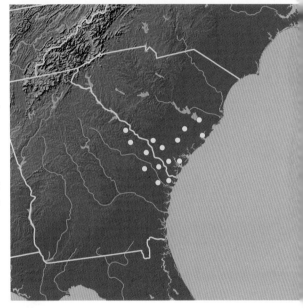

South Carolina Slimy Salamander

Plethodon vehiculum (Cooper, 1860)

Western Red-backed Salamander

Western Red-backed Salamanders range through out much of western Washington and Oregon west of the Cascades Range crest south to the headwaters of the North Umpqua River and northward into the lower Fraser Valley and southern Coast Mountains of southwest British Columbia. They also occur throughout Vancouver Island and adjacent coastal islands in the Strait of Georgia and Puget Sound. There is also a record for them on Moresby Island. Western Red-backed Salamanders occur at elevations from sea level to about 1250 m.

Western Red-backed Salamanders are widespread and occur in a variety of habitats, including rocky talus slopes and rocky seeps, on the sides of springs and small streams, and on the floor of cool, shaded forests. They are sometimes extremely abundant beneath logs, bark, stones, and moss. Hatchlings can be numerous on the forest floor in late autumn and early winter. Juveniles tend to use smaller cover objects, such as leaf litter and stones, as compared with adults. During drier conditions, juvenile Western Red-backed Salamanders tend to be more active on the forest floor than are adults.

Western Red-backed Salamanders are not listed in either Canada or the U.S.

ORIGINAL ACCOUNT by R. Bruce Bury; photograph by Kristina Ovaska.

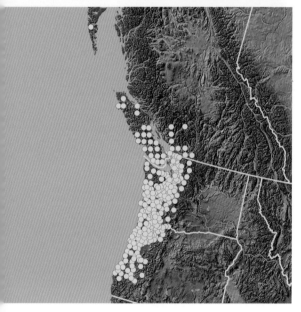

Western Red-backed Salamander

Plethodon ventralis Highton, 1997

Southern Zigzag Salamander

Southern Zigzag Salamanders occur in isolated sites scattered along the southern Appalachians at elevations less than 580 m from northern Mississippi to central Kentucky and southwestern Virginia, including sites in northern Alabama and Georgia, eastern Tennessee, and western North Carolina.

Southern Zigzag Salamanders are found in flat, moist areas of forest, including talus on lower mountain slopes. Southern Zigzag Salamanders are rarely found on the surface during the summer but it is not unusual to find individuals under rocks and logs on mild days during the winter.

Southern Zigzag Salamanders are listed as a Species of Special Concern in North Carolina.

ORIGINAL ACCOUNT by David A. Beamer and Michael J. Lannoo; photograph by Travis Brown.

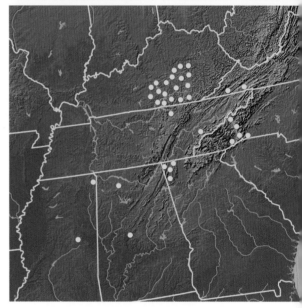

Southern Zigzag Salamander

Plethodon virginia Highton, 1999

Shenandoah Mountain Salamander

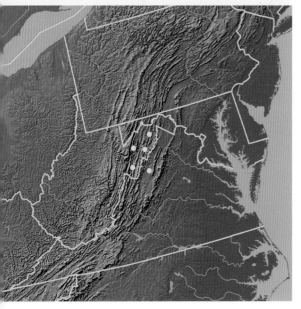

Shenandoah Mountain Salamander

Shenandoah Mountain Salamanders occur in eastern West Virginia and adjacent northern Virginia, including the Shenandoah, South Branch, and Nathaniel mountains, and west to the South Fork of the South Branch of the Potomac River in West Virginia at elevations between 1100 and 1200 m.

Shenandoah Mountain Salamanders are found on ridges characterized by deep soil that supports vegetation comprised of white oak, pink honeysuckle, and late low blueberry, and on slopes characterized by shallow, rocky soil on which grow chestnut oak, and red maple, late low blueberry bushes, witch hazel, and mountain laurel. Farther down the slopes, Shenandoah Mountain Salamanders occur in places where the soil is rocky, with exposed bedrock and covered by chestnut oak, red oak, bitternut hickory, black gum, and witch hazel.

Shenandoah Mountain Salamanders do not have conservation status in either Virginia or West Virginia, the two states in which they occur.

ORIGINAL ACCOUNT by David A. Beamer and Michael J. Lannoo; photograph by Gary Nafis.

Plethodon websteri Highton, 1979

Webster's Salamander

Webster's Salamanders occur in a small number of scattered sites from eastern Louisiana, through Mississippi, Alabama, and Georgia, to western South Carolina.

Webster's Salamanders are found in humid mixed forest bordering rocky, lower-order streams, in limestone ravines, in small stream valleys and in pastures taken over by second-growth forest. The predominant tree species in these forests include sugar maple, pignut hickory, red oak, yellow poplar, sweetgum, and slippery elm. In such places, Webster's Salamanders may be found under rocks and decaying logs.

Webster's Salamanders are listed as an Endangered species in South Carolina and as a Species of Special Concern in Louisiana.

ORIGINAL ACCOUNT by David A. Beamer and Michael J. Lannoo; photograph by Ken Dodd.

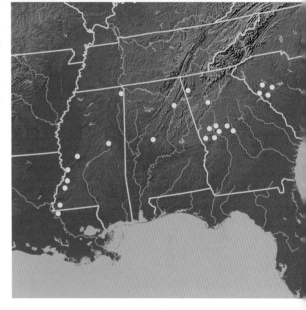

Webster's Salamander

Plethodon wehrlei Fowler and Dunn, 1917

Wehrle's Salamander

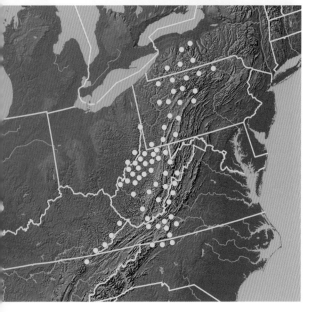

Wehrle's Salamanders range through the Appalachians from the Allegheny Mountains of southwestern New York, western Pennsylvania, and eastern Ohio through to, western North Carolina, extreme eastern Kentucky and adjacent Tennessee.

The preferred habitat of Wehrle's Salamanders varies from mixed deciduous forests in the Allegheny Plateau and Cumberland Plateau to coniferous forests at high elevations in the Allegheny Mountains. Wehrle's Salamanders seek cover in these habitats under logs, rocks, and leaves. They also occur in caves.

Although Wehrle's Salamanders may, in places, be the most abundant salamanders in their habitat, deforestation is a serious threat. They have not been recorded in Ohio since the 1930's.

Wehrle's Salamanders are listed as In Need of Conservation in Maryland, In Need of Management in West Virginia, and Threatened in North Carolina.

ORIGINAL ACCOUNT by Thomas K. Pauley and Mark B. Watson; photograph by Michael Graziano.

Wehrle's Salamander

Plethodon welleri Walker, 1931

Weller's Salamander

Weller's Salamanders inhabit high-elevation regions, usually above 1500 m, in the Blue Ridge Mountains of southwestern Virginia, extreme northeastern Tennessee, and northwestern North Carolina, including areas along the Unaka Mountain ridges and Grandfather Mountain.

Weller's Salamanders are generally found in spruce–fir forests, beneath logs, stones, and flakes of rock covering high slopes. They are occasionally found at lower elevations in coves associated with limestone or in mixed deciduous forests below spruce–fir forests.

Populations of Weller's Salamanders, being restricted to high mountain peaks, are considerably isolated from one another, but several of these mountain areas are on state or federal property and enjoy some degree of protection. Spruce–fir forest die-offs may constitute a long-term environmental threat to Weller's Salamanders.

Weller's Salamanders are listed as a Species of Special Concern in North Carolina and as Wildlife in Need of Management in Tennessee.

ORIGINAL ACCOUNT by David A. Beamer and Michael J. Lannoo; photograph by Michael Graziano.

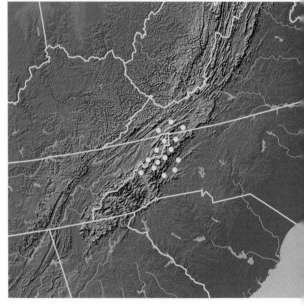

Weller's Salamander

Plethodon yonahlossee Dunn, 1917

Yonahlossee Salamander

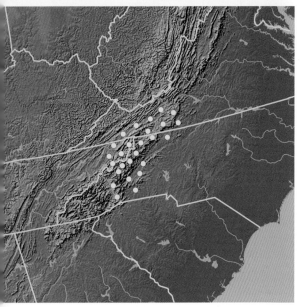

Yonahlossee Salamander

Yonahlossee Salamanders are found in the Blue Ridge Mountains of western North Carolina, northeastern Tennessee, and southwestern Virginia at elevations between about 440 and 1740 m.

Yonahlossee Salamanders occur in both virgin and second-growth deciduous forest, where they live in exposed rock-covered road embankments, in talus slopes and even in open pasture. In these settings, they are often found in and under rotting logs, under bark on the ground or still on its log, and under stones, though they appear to have a particular preference for old windfalls that have shed most of their bark and larger fallen logs lying amid a thick layer of accumulated fallen leaves. Yonahlossee Salamanders emerge from their burrows at night to forage. Juveniles are most active for about 1 hour after sunset, while adult activity peaks 1 or 2 hours later.

As Yonahlossee Salamanders are usually found from intermediate to high elevations, patches of suitable habitat may be isolated from each other. Though they can withstand minor disturbances associated with logging operations, aggressive forestry practices may seriously affect populations of Yonahlossee Salamanders.

Yonahlossee Salamanders are not listed in any of the states in which they are found.

ORIGINAL ACCOUNT by David A. Beamer and Michael J. Lannoo; photograph by Michael Graziano.

Pseudotriton montanus Baird, 1849

Mud Salamander

Mud Salamanders are found from extreme southeastern Louisiana east to the Atlantic Coast, north to southern New Jersey, and westward to the Illinois border. Predominantly a lowland species, Mud Salamanders are absent from the higher elevations of the Appalachian Mountains.

Mud Salamanders inhabit lowland seeps, marshes, muddy springs and streams, and swampy pools and ponds. They live in burrows that are usually located either near the water or a short distance away. The burrows, which the salamanders construct themselves, contain vertical channels from the surface to the water table below. Adult Mud Salamanders are largely subterranean and reside in their burrows with their heads near the surface. Little is known about their abundance as they are rarely seen at the surface.

Mud Salamanders are listed as Rare and on the Watch List in Maryland, as a Species of Special Concern in Louisiana and South Carolina, as Rare in Ohio and West Virginia, as Threatened in New Jersey, and as Endangered in Pennsylvania.

ORIGINAL ACCOUNT by Todd W. Hunsinger; photograph by Mike Redmer.

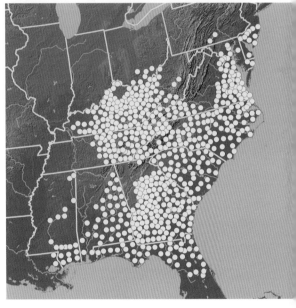

Mud Salamander

Pseudotriton ruber (Latreille, 1801)

Red Salamander

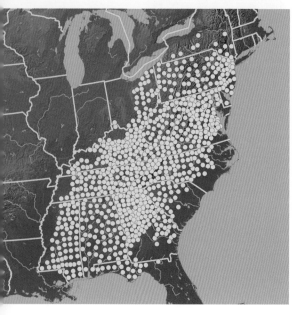

Red Salamander

Red Salamanders are found from the Mohawk and Hudson rivers in southeast New York, through eastern Ohio to western Kentucky and east of the Mississippi river to the Gulf Coast from Lake Borgne in Louisiana to the Suwanee River in northern Florida. They are absent, though, from much of the southern Atlantic Coast and from peninsular Florida.

Red Salamanders inhabit slow-moving headwater springs and seeps in wooded lowland and upland areas, tamarack wetlands and even deep, acidic lakes. Adults are found under rocks, logs, or mats of sphagnum moss in wooded ravines, swamps, open fields, and meadows, and they use burrows that connect to watercourses.

Red Salamanders are known to have vanished from several places where they historically occurred and many other populations appear to be less robust than they once were.

Red Salamanders are listed as Endangered in Indiana, as a Species of Special Concern in Louisiana, and as Protected in New Jersey.

ORIGINAL ACCOUNT by Todd W. Hunsinger; photograph by Mike Redmer.

Stereochilus marginatus (Hallowell, 1856)

Many-lined Salamander

Many-lined Salamanders inhabit the Atlantic Coastal Plain from approximately the James River in Virginia south to northeastern Florida.

Many-lined Salamanders live in the shallow, acid waters of Lower Coastal Plain swampy streams, gum and cypress swamps, woodland ponds, borrow pits, canals, and drainage ditches. They are usually aquatic, especially in permanent water, but occasionally are found on land in damp situations under logs, among dead leaves and detritus or in and under sphagnum mats.

Wetlands along the Atlantic Coastal Plain where Many-lined Salamanders are found are rapidly disappearing or being substantially altered. Conversion and destruction of wetlands is likely a common primary threat to the long-term persistence of Many-lined Salamanders.

Many-lined Salamanders are not listed either federally or at the state level in the U.S.

ORIGINAL ACCOUNT by Travis J. Ryan; photograph by Michael Graziano.

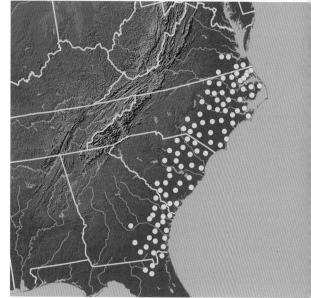

Many-lined Salamander

Urspelerpes brucei Camp, Peterman, Milanovich, Lamb, and Wake 2009[16]

Patch-nosed Salamander

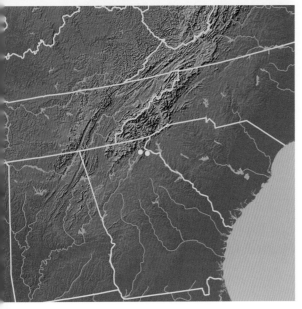

Patch-nosed Salamanders are known from the base of the Blue Ridge escarpment in northeast Georgia.

Patch-nosed Salamanders inhabit shallow, first and second-order streams flowing over sand and gravel through mature, beech–oak dominated, mixed deciduous forests. They are found within the stream banks in association with moist, but not inundated, microhabitats. Larval Patch-nosed Salamanders, though, have a proclivity for leaf packs within the stream. The majority of their growth appears to occur during the larval stage, and sexual maturity is approached in newly metamorphosed individuals.

Within their extremely restricted distribution and limited habitat, Patch-nosed Salamanders appear to be both uncommon and hard to detect.

Patch-nosed Salamanders receive no legal protection either federally or at the state level in the U.S.

ACCOUNT by William Peterman; photograph by William Peterman.

Patch-nosed Salamander

Necturus alabamensis Viosca, 1937

Black Warrior River Waterdog

Black Warrior River Waterdogs are confined to medium and large streams above the Fall Line in north central Alabama, including the upper Black Warrior River, Coosa River and Sipsey River drainages, and the adjacent Tennessee River watershed.

Black Warrior River Waterdogs are wholly aquatic and are associated with silt-free rivers with clay substrates. The presence of leaf beds within these rivers is important, as they are rich with invertebrate faunas and larval Black Warrior River Waterdogs typically are found submerged within them.

Black Warrior River Waterdogs appear to have declined in abundance, and their local distribution, even within the best habitat, appears to be patchy.

Although they are currently a candidate species for U.S. federal listing and are rated as Endangered by the IUCN, Black Warrior River Waterdogs receive no legal protection at the state level in Alabama.

ORIGINAL ACCOUNT by Mark A. Bailey; photograph by Mike Redmer.

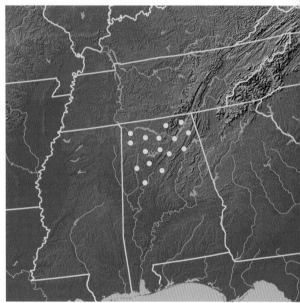

Black Warrior River Waterdog

Necturus beyeri Viosca, 1937

Gulf Coast Waterdog

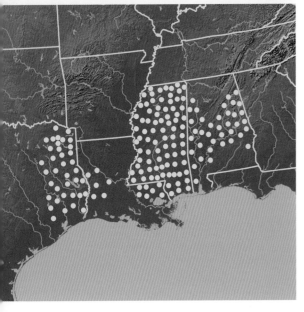

Gulf Coast Waterdog

Gulf Coast Waterdogs occur in two distinct regions on either side of the lower Mississippi River. Western populations of Gulf Coast Waterdogs are found in eastern Texas and west central Louisiana in rivers draining into Galveston Bay and Sabine Lake, whereas eastern populations are found in Mississippi and central Alabama in rivers draining into the Gulf of Mexico along the coast from Lake Borgne to Mobile Bay.

Gulf Coast Waterdogs are found in sandy, spring-fed streams rather than the turbid, sluggish waters characteristic of bayous, rivers, and lakes of the lower Mississippi River system. Most adult Gulf Coast Waterdogs are found in slow-moving sections of streams under large objects such as logs, flood debris, and other obstructions. Some individuals live in stream bank burrows. Juvenile Gulf Coast Waterdogs inhabit bottom debris, especially leaf litter, where currents are slow and prey are plentiful.

Populations of Gulf Coast Waterdogs likely have been reduced because of increased siltation and pollution in the rivers they inhabit.

Gulf Coast Waterdogs are not listed in the U.S. under any state or federal laws or regulations.

ORIGINAL ACCOUNT by Craig Guyer; photograph by John Clare.

Necturus lewisi Brimley, 1924

Neuse River Waterdog

Neuse River Waterdogs are known only from the Neuse and Tar river systems in the Piedmont and Atlantic Coastal Plain regions of North Carolina.

Neuse River Waterdogs are found in rivers ranging from large headwater streams to coastal streams up to the point of saltwater intrusion. They can be found associated with leaf beds in wide, shallow backwaters off the main current, where substrates are sandy, muddy, or composed of clay. Both juvenile and adult Neuse River Waterdogs construct retreats under submerged rocks or other cover objects, with entrances on the downstream side. At night, they become active and will leave their cover.

Populations of Neuse River Waterdogs have been lost from severely polluted streams.

Neuse River Waterdogs are considered a Species of Special Concern in North Carolina.

ORIGINAL ACCOUNT by Alvin L. Braswell; photograph by Ken Dodd.

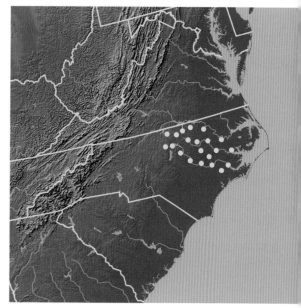

Neuse River Waterdog

Necturus maculosus (Rafinesque, 1818)

Mudpuppy

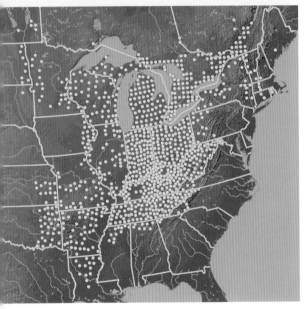

Mudpuppy

Mudpuppies are distributed throughout much of east central North America from southern Manitoba, Ontario, and Québec south to Louisiana, including the Great Lakes and St. Lawrence River drainage, the Red, Rainy, Mississippi and Hudson river systems, Lake Champlain, and the rivers of Alabama, Georgia, and North Carolina. Mudpuppies have been reported at depths down to 27 m in Lake Michigan.

Mudpuppies are wholly aquatic and are found in lakes, reservoirs, canals, ditches, and streams, preferably where there are flat rock slabs, submerged logs, crayfish burrows, undercut banks, or tree roots available for them to hide under. Adult Mudpuppies particularly prefer well-aerated water downstream or to the sides of riffles. Larvae and juvenile Mudpuppies up to 4 years of age are found in greater numbers in pools where silt and organic debris have accumulated to a thickness of several centimeters or more.

Chemical water pollutants and heavy siltation that have reduced habitat quality in many regions have contributed to declines in the abundance of Mudpuppies. Mudpuppies are often used for fish bait or are incidentally hooked by ice fishermen and often left to die on the ground or on the ice.

Mudpuppies are considered Endangered/ Extirpated by Maryland, Threatened in Iowa, and a Species of Special Concern in Indiana and North Carolina.

ORIGINAL ACCOUNT by Timothy O. Matson; photograph by Mike Redmer.

Necturus punctatus (Gibbes, 1850)

Dwarf Waterdog

Dwarf Waterdogs are found in Atlantic Coastal Plain streams from the Chowan River in southeastern Virginia through to the Ocmulgee–Altamaha River system in southeastern Georgia.

Dwarf Waterdogs are permanently aquatic and are usually found in the slower regions of small to medium-sized streams, including side ditches, with muddy or sandy banks or bottoms. Dwarf Waterdogs are most common in deeper sections of these streams, where there is an accumulation of mud, silt, and leaves. They are rarely found in the main channels of rivers. Juvenile Dwarf Waterdogs apparently prefer shallower water than adults and are most common in mats of leaves.

Despite stream pollution, Dwarf Waterdogs still occur widely throughout their range.

Dwarf Waterdogs are not listed at either the state or federal level in the U.S.

ORIGINAL ACCOUNT by Harold A. Dundee; photograph by Michael Graziano.

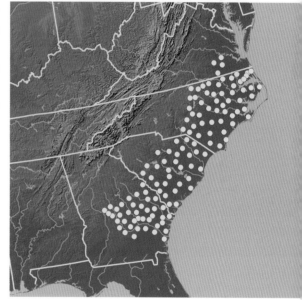

Dwarf Waterdog

Rhyacotriton cascadae Good and Wake, 1992

Cascade Torrent Salamander

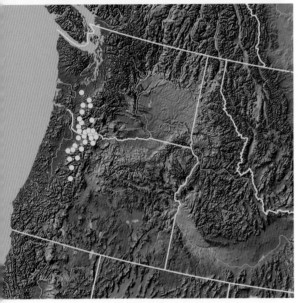

Cascade Torrent Salamander

Cascade Torrent Salamanders are restricted to the west slope of the Cascades Range from the west bank of the Skookumchuck River in western Washington south to the Middle Fork of the Willamette River in central western Oregon, up to elevations at which heavy snow accumulates during winter.

Forests bordering streams that harbor Cascade Torrent Salamanders usually have a good leaf canopy, abundant understory vegetation, much moss, and a thick leaf mat along the stream banks. Cascade Torrent Salamanders are frequently found in the rock rubble of stream banks, fissures in stream banks, underground water courses, fissures in stream heads, and cracks in wet cliff faces.

The conversion of mature and old-growth forests to young stands because of timber harvesting may result in increased temperatures and sedimentation in the streams inhabited by Cascade Torrent Salamanders, to their detriment.

Cascade Torrent Salamanders have Sensitive status in both Oregon and Washington.

ORIGINAL ACCOUNT by Marc P. Hayes; photograph by William Leonard.

Rhyacotriton kezeri Good and Wake, 1992

Columbia Torrent Salamander

Columbia Torrent Salamanders are restricted to coastal and near-coastal regions of the Coast Range from the Little Nestucca River in northwestern Oregon north to the Chehalis River in southwestern Washington. They occur at elevations from near sea level to the highest in the region—about 1000 m.

Columbia Torrent Salamanders are characteristically found in and along cold, permanent streams with small, water-washed or moss-covered rocks and rock rubble. They usually select sites in seeps and small, trickling tributary streams where the movement of water is relatively slow. They habitually remain in contact with water or, at least, a wet, saturated substrate.

There is concern that habitat quality for Columbia Torrent Salamanders is degraded by increased temperatures and sedimentation following timber harvest.

Columbia Torrent Salamanders are listed as Sensitive in both Oregon and Washington.

ORIGINAL ACCOUNT by Marc P. Hayes and Timothy Quinn; photograph by William Leonard.

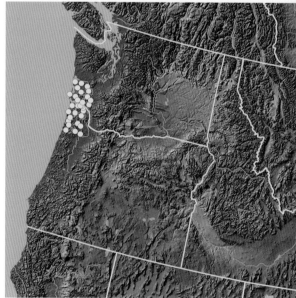

Columbia Torrent Salamander

Rhyacotriton olympicus (Gaige, 1917)

Olympic Torrent Salamander

Olympic Torrent Salamanders are limited to the region of the Olympic Peninsula in Washington as far south as the Chehalis River.

Olympic Torrent Salamanders are nearly always seen in or very near cold, clear streams, seepages, or waterfalls, typically in or near the splash zone, where a thin film of water runs between or under rocks. They are less often found where conditions are drier and warmer on the east side of the Olympic Mountains because of the significant rain shadow effect.

Timber harvesting that potentially leads to the degradation of the streams, seeps, and springs occupied by Olympic Torrent Salamanders may be a threat to their continued well-being.

Olympic Torrent Salamanders are listed as Sensitive in Washington.

ORIGINAL ACCOUNT by Marc P. Hayes and Lawrence L.C. Jones; photograph by William Leonard.

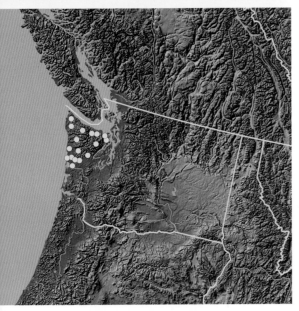

Olympic Torrent Salamander

Rhyacotriton variegatus Stebbins and Lowe, 1951

Southern Torrent Salamander

Southern Torrent Salamanders are distributed in the Coast Range of western Oregon and northern California from the Little Nestucca River south to Dark Gulch. Southern Torrent Salamanders also occur in the north Umpqua River drainage of the interior southern Cascades Range of Oregon and from the upper McCloud River drainage in California.

Southern Torrent Salamanders are closely linked to cold, headwater stream habitats in wet, mature to old-growth forests. Adult Southern Torrent Salamanders, while occasionally found in adjacent moist vegetation, are usually found in contact with cold water in springs, seeps and streams with shallow, slow flows of water over unsorted rock or rock rubble substrates. The coarse substrate interstices of the streambed are used for cover by both larval and adult Southern Torrent Salamanders.

Fine sediments of sand and gravel that fill in rocky streambeds have a negative impact on Southern Torrent Salamanders. It is likely that populations of these salamanders may be extirpated where logging and related land-management practices lead to such degradation of streams.

Southern Torrent Salamanders are not listed by either the U.S. Fish and Wildlife Service or the California Department of Fish and Game Commission.

ORIGINAL ACCOUNT by Hartwell H. Welsh Jr. and Nancy E. Karraker; photograph by Dante Fenolio.

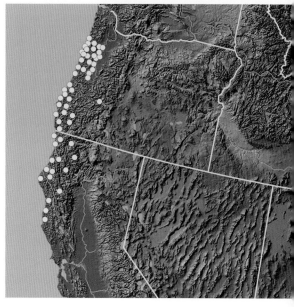

Southern Torrent Salamander

Notophthalmus meridionalis (Cope, 1880)

Black-spotted Newt

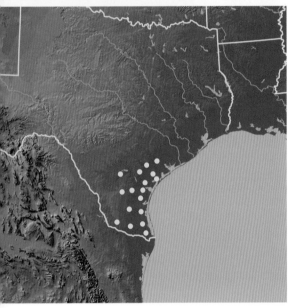

Black-spotted Newt

Black-spotted Newts inhabit the southern Gulf Coastal Plain of Texas south into northern Veracruz in Mexico (not shown).

Black-spotted Newts live where deep, poorly drained, clay soils with slow permeability allow for the formation of ephemeral ponds or wetlands during periods of heavy rain. Adults generally are to be found in, or near, such breeding ponds in the presence of intact Tamaulipan thorn forest. Juvenile Black-spotted Newts may remain aquatic until reproductively mature, unless their pond dries up or high temperatures cause them to seek cover on land.

Continued land clearing for agriculture and urban development poses a considerable threat to Black-spotted Newts, as over 95% of the original Tamaulipan brushland of the lower Rio Grande valley has now been lost.

The Texas Parks and Wildlife Department lists Black-spotted Newts as Endangered, but they have no U.S. federal listing status.

ORIGINAL ACCOUNT by Kelly J. Irwin and Frank W. Judd; photograph by John Clare.

Notophthalmus perstriatus Bishop, 1941

Striped Newt

Striped Newts are found on the Dougherty Plain and Atlantic Coastal Plain of southern Georgia south to central Florida.

Striped Newts inhabit longleaf pine savanna characterized by a rich groundcover of grasses and forbs, with a turkey oak midstory. Adult Striped Newts sometimes are found up to 700 m from the nearest breeding pond. Presumably, they stay under cover in subterranean refugia and, occasionally, they are found under fallen logs.

Striped Newts have declined substantially throughout their range because of habitat loss and degradation due to fire-suppression and silvicultural practices, pond drainage, and fish introductions.

Although striped Newts are not listed under U.S. statutes, the U.S. Fish and Wildlife Service is concerned about their conservation status and considers the species as Under Review. Striped Newts are listed as Rare in Georgia, the Florida Natural Areas Inventory considers them as Imperiled in Florida, and the Florida Committee on Rare and Endangered Plants and Animals lists them as Rare.

ORIGINAL ACCOUNT by C. Kenneth Dodd Jr., D. Bruce Means, and Steve A. Johnson; photograph by Michael Graziano.

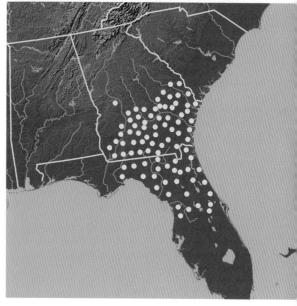

Striped Newt

Notophthalmus viridescens (Rafinesque, 1820)

Eastern Newt

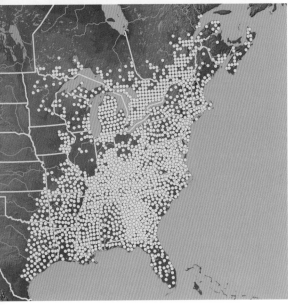

Eastern Newt

Eastern Newts are found throughout eastern North America from the Rainy River district of western Ontario, east to the Canadian Maritime Provinces, and south to the Gulf of Mexico from east Texas to the tip of the Florida Peninsula.

Eastern Newts are aquatic as adults and inhabit pools, ponds, wetlands, sloughs, canals, and quiet areas of streams in historically forested areas. Sites with submergent and emergent vegetation often harbor large populations. During the winter, aquatic adult Eastern Newts will sometimes congregate in ice-free areas of ponds where temperatures are at 5 to 6°C. At times, as when their ponds are at risk of drying up, adults will leave them for protected terrestrial sites and will hide under rotten logs and vegetation clumps to avoid desiccation and heat stress. Juvenile Eastern Newts, known as efts, are terrestrial and are usually found in wooded areas. Efts will move about on rainy or humid days and nights when the ground is moist, but during dry periods, they will seek cover in leaf litter or under logs and other objects.

Eastern Newts are good colonizers and will readily use new beaver ponds and farm ponds. They also benefit from the reforestation that is occurring throughout much of their range.

Eastern Newts are listed as Threatened in Iowa and in Kansas.

ORIGINAL ACCOUNT by Todd W. Hunsinger and Michael J. Lannoo; photograph by Mike Redmer.

Taricha granulosa (Skilton, 1849)

Rough-Skinned Newt

Rough-skinned Newts range from the Santa Cruz Mountains of California north up the Pacific Coast as far as Admiralty Island in Alaska, including Vancouver Island, Prince of Wales Island, and other inshore islands in Puget Sound, the Strait of Georgia, and Clarence Strait. They extend inland to Lillooet up the Fraser River Valley in British Columbia, to the east slope of the Cascades Range in Washington, and along the west slope of the Sierra Nevada in northern California. Isolated and extremely small populations near Moscow, Idaho, and Thompson Falls, Montana, may be the result of introductions. Rough-skinned Newts occur at elevations from sea level to about 2800 m.

Rough-skinned Newts are found in a variety of habitats, such as coniferous forests, redwood forests, oak woodland, farmlands, and grassland. Terrestrial adults spend much of their time underground in burrows or beneath logs, bark, or boards. However, adult Rough-skinned Newts can be highly aquatic and inhabit lakes, ponds, roadside ditches, and slow-moving portions of streams with surrounding vegetation. In southern Vancouver Island, adult males remain mostly aquatic throughout the year, whereas females leave the pond with the onset of fall rains, overwinter on land, and migrate in spring to breeding ponds. At a few, mostly high-elevation, localities in the Cascades Range, gilled adults of

Rough-Skinned Newt

both sexes may be permanently aquatic. In northern California, adult Rough-skinned Newts living in streams usually remain aquatic all year unless forced to leave when water flow rises during winter floods.

Although Rough-skinned Newts are protected under the provincial Wildlife Act in British Columbia, they are not listed at the federal level in Canada or under any state or federal regulations in the U.S. They are considered an Exotic Species in Idaho.

ORIGINAL ACCOUNT by Sharyn B. Marks and Darrin Doyle; photograph by William Leonard.

Taricha rivularis (Twitty, 1935)

Red-bellied Newt

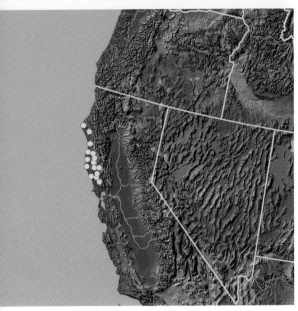

Red-bellied Newt

Red-bellied Newts occur in coastal California from the Estero Americano north to the Eel River in the western versant of the Coast Range at elevations between 150 and 450 m.

Red-bellied Newts are found on steep, heavily wooded, north-facing slopes that are littered with many fallen trees and branches that provide them with cover. Adult Red-bellied Newts migrate from such terrestrial habitats seasonally to nearby streams and rivers for breeding. They apparently do not use ponds or other standing water habitats for breeding.

Red-bellied Newts have a limited and somewhat spotty geographic distribution, and human population pressure has intensified considerably over much of their range. Conversion of native forests and grasslands to vineyards and subdivisions likely poses a serious threat to Red-bellied Newts, as does degradation of aquatic habitats. Finally, increased vehicular traffic associated with housing subdivisions undoubtedly has resulted in increased mortality of Red-bellied Newts.

Red-bellied Newts are a species of Special Concern in California.

ORIGINAL ACCOUNT by Sharyn B. Marks and Darrin Doyle; photograph by Gary Nafis.

Taricha sierrae (Twitty, 1942)[17]

Sierra Newt

Sierra Newts occur in California along the western slopes of the Sierra Nevada from near to Mount Shasta south to the Kaweah River.

Sierra Newts inhabit forested hill regions dominated by conifers, especially gray pine and ponderosa pine. In January and February, adult Sierra Newts migrate to aquatic breeding sites, which range from swift-flowing streams to still-water farm ponds, lakes, and ditches. During the dry summer months, they spend their time in moist, terrestrial habitats, preferentially residing under woody debris or in animal burrows. They emerge again to be active on the surface with the onset of autumn rains.

Sierra Newts appear to be able to adapt to fluctuating conditions in streams. There is, however, a possible threat to aquatic larval Sierra Newts from introduced fishes, such as stocked trout, and introduced American Bullfrogs have also been observed to eat both juveniles and adults.

Sierra Newts are considered to be a Species of Special Concern by the state of California.

ACCOUNT by David M. Green; photograph by Gary Nafis.

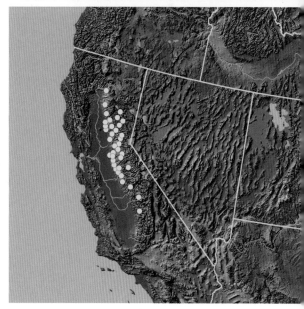

Sierra Newt

Taricha torosa (Rathke, 1833)

California Newt

California Newt

In the Coast Ranges of California, California Newts occur at elevations from sea level to 2000 m from the vicinity of Round Valley south as far as Boulder Creek in the Peninsular Ranges of extreme southwestern California. California Newts also occur on the western side of the southern Sierra Nevada from the Kaweah River south to the vicinity of the Kern River.

California Newts can be found in a variety of habitats, including mountainous or rolling woodland, grassland, chaparral, or forest, so long as there are permanent or semipermanent pools or ponds in which they may breed.

California Newts are badly affected by breeding-site degradation, destruction of summer aestivation sites and migration routes, road mortality, large-scale commercial exploitation, and altered sedimentation dynamics in stream pools resulting from wildfires. Exotic crayfish and mosquitofish present in many places in southern California may exclude California Newts from their breeding sites. In consequence, many populations of California newts have declined or even been extirpated.

California Newts south of the Salinas River are considered by the California Department of Fish and Game to be Species of Special Concern.

ORIGINAL ACCOUNT by Shawn R. Kuchta; photograph by Mike Redmer.

Pseudobranchus axanthus Netting and Goin, 1942

Southern Dwarf Siren

Southern Dwarf Sirens are restricted to peninsular Florida south of the Suwannee River in the west and the Nassau River in the east.

Southern Dwarf Sirens are found in heavily vegetated marshes and shallow lakes. They may be abundant in floating mats of vegetation or in mucky shoreline deposits.

The current distribution and abundance of Southern Dwarf Sirens have undoubtedly declined as wetlands in peninsular Florida have been reduced through drainage of surface waters associated with residential, agricultural, and silvicultural development.

Southern Dwarf Sirens are not listed under any Florida or U.S. federal laws or regulations.

ORIGINAL ACCOUNT by Paul E. Moler; photograph by Todd Pierson.

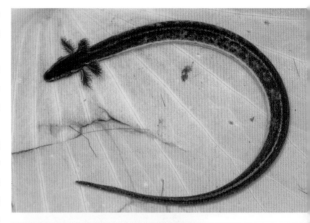

Southern Dwarf Siren

Pseudobranchus striatus (LeConte, 1824)

Northern Dwarf Siren

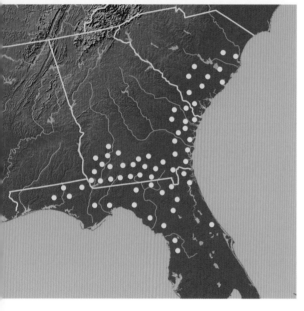

Northern Dwarf Siren

Northern Dwarf Sirens occur in the lower Gulf and Atlantic Coastal Plains from the Pee Dee River in South Carolina south to central peninsular Florida and west to Choctawhatchee Bay.

Northern Dwarf Sirens are wholly aquatic and are most often associated with cypress or gum ponds and other shallow, acidic wetlands of the flatwoods. They are not normally found among water hyacinths, which are typically absent from such sites. Northern Dwarf Sirens usually inhabit decaying bottom vegetation and the soft, mucky soils of pond margins, though they will also be found among floating mats of vegetation.

Populations of Northern Dwarf Sirens have certainly been lost as wetland habitats have been reduced through drainage of surface waters associated with residential, agricultural, and silvicultural development.

Northern Dwarf Sirens are considered Threatened in South Carolina.

ORIGINAL ACCOUNT by Paul E. Moler; photograph by Pierson Hill.

Siren intermedia Barnes, 1826[18]

Lesser Siren

Lesser Sirens inhabit the Atlantic and Gulf Coastal Plains from Virginia to southern Texas and adjacent Mexico (not shown), including most of peninsular Florida, and range north up the Mississippi Valley to central Illinois, and up the valley of the Wabash River through western Indiana to southwestern Michigan. It is possible that their range has been expanded by fish stocking.

Lesser Sirens live in the shallow, warm, quiet waters of ponds, sloughs, marshes, swamps, ditches, canals, and sluggish, vegetation-choked creeks where aquatic vegetation is plentiful. Temporary floodplain pools and shallow, heavily vegetated sections of ponds with deep sediments provide them with burrowing sites. Juvenile Lesser Sirens are most active at night and live in burrows or in thick mats of aquatic vegetation, where they forage on small invertebrates.

Populations of Lesser Sirens are becoming increasingly isolated from one another by flood-control programs and wetland drainage. They are possibly extirpated from the Great Lakes drainage basin.

Lesser Sirens are considered Threatened in Michigan and as Rare and Endangered in Kentucky.

ORIGINAL ACCOUNT by William T. Leja; photograph by Mike Redmer.

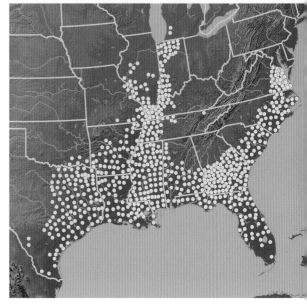

Lesser Siren

Siren lacertina Linnaeus, 1766

Greater Siren

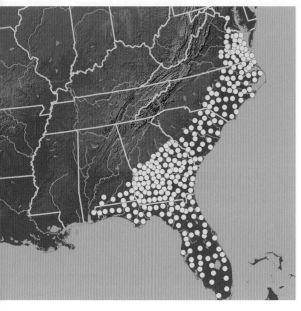

Greater Siren

Greater Sirens are found in the Atlantic and Gulf Coastal Plains from the Potomac River through to extreme southern Florida and west to Mobile Bay in Alabama.

Greater Sirens are permanently aquatic and are found in muddy and weed-choked ditches, swamps, and ponds, as well as large lakes and streams. When faced with desiccating conditions, Greater Sirens burrow into the muddy bottom. Juvenile Greater Sirens are often seen amid water hyacinth roots or other heavy vegetation.

Greater Sirens have been considered common in the more central portions of their range and from locally abundant to rare in more peripheral sites.

In Maryland, Greater Sirens are listed as Endangered/Extirpated and are afforded legal protection.

ORIGINAL ACCOUNT by Russ Hendricks; photograph by Todd Pierson.

MAPPING NORTH AMERICAN AMPHIBIANS

MAKING THE MAPS

The background for all the maps in this book is composed of several layers. The terrain is a color-shaded relief image at 1-kilometer resolution obtained from the *National Atlas of the United States* (http://nationalatlas.gov/mld/shdrlfi.html). Overlaid on that image are the political boundaries and hydrography data from the *North American Atlas* (http://nationalatlas .gov/atlasftp-na.html), a cooperative intergovernmental venture by Natural Resources Canada, the U.S. Geological Survey, and the Mexican Instituto Nacional de Estadística Geografía e Informática. We then used ESRI® ArcMap™ versions 9.3.1 and 10 Geographical Information Systems software to plot our distribution data as dots on top of these background layers to make the maps.[19]

We compiled amphibian distribution data from many sources, including museum records, provincial conservation data center records, agency records, amphibian survey records, government reports, atlases, field notes, published documents and data from institutions participating in online resources such as HerpNet. For U.S. species distribution information on our maps, we relied on version 3.0 of the U.S. Geological Survey National Amphibian Atlas (http://www.pwrc.usgs.gov/naa). This dataset is a revised and updated version of the dataset used for the 2005 book *Amphibian Declines: The Conservation Status of United States Species*. The original dataset was sent to the U.S. Geological Survey Patuxent Wildlife Research Center, where it has been maintained and periodically updated since 2005.

For most of the U.S., our maps show county-level distributions, such that each dot is the centroid of a county where the species occurs. For the five western states of Arizona, California, Nevada, Oregon, and Washington that have counties with large land areas, we used centroids of Minor Civil Divisions instead of counties. Minor Civil Divisions are sub-county units defined by the U.S. Census Bureau. We obtained digital files of county and Minor Civil Division boundaries from the U.S. Census Bureau's Topologically Integrated Geographic Encoding and Referencing (TIGER) system. We merged the subcounty units

317

of the five western states with county units for the remaining states to create a single set of map units. Species distributions were tracked by island for the Hawaiian Islands and Channel Islands of California. The map units were then linked with a relational amphibian species database through a common data field that provided a unique identifier for each unit based on U.S. Federal Information Processing Standards (FIPS). For Canada, where provincial districts are too large to be comparable to U.S. counties, we created a grid in which each grid cell was 3500 square meters, designed to be roughly comparable in size to U.S. counties. We then mapped Canadian species occurrence records to the grid, where each dot is the centroid of a grid cell where at least one occurrence record was obtained for the species.

DEALING WITH TAXONOMIC CHANGES AND NEW INFORMATION

In this book, we use the currently accepted list of species recognized by the 7th edition of *Scientific and Standard English Names of Amphibians and Reptiles of North America North of Mexico*, edited by Brian Crother and published in 2012. Though we have neither mapped nor discussed any subspecies mentioned in the list, we have written about and mapped every species, with just two exceptions. Recognizing the distinction of the unisexual populations of salamanders, genus *Ambystoma,* from the Jefferson Salamander, Blue-spotted Salamander, and other species to which they are related, we have mapped their distribution separately from these species. As shown by Ke Bi and James P. Bogart (*BMC Evolutionary Biology* 10:238, 2010), these salamanders constitute a truly bizarre genetic complex, ancient in origin, of polyploid individuals containing a mixtures of genomes. On the other hand, we have made only one map for the Gray Treefrog and Cope's Gray Treefrog. Although Alisha K. Holloway and coauthors (*American Naturalist* 167:E88–E101, 2006) have provided a blob map showing the ranges of the two species based on genetic information, most of the occurrence records for the two species can be trusted only to document the presence of a "gray treefrog" of one species or the other because of the impossibility of telling the two species apart morphologically.

Taxonomic changes inevitably confound the art of making range maps. Advances in our understanding of the genetic diversity of amphibians have, in many cases, led to discoveries of cryptic species that were not known when the books edited by Green and Lannoo were published. Where once we may have thought that a particular species occupied an extensive geographic range, we now may have to recognize that two or more species actually occur. Erica J. Crespi and coauthors (*Herpetologica,* 66: 283–295, 2010) provided explicit information for delineating the respective ranges of Pigmy Salamanders (*Desmognathus wrighti*) and Northern Pigmy Salamanders (*D. organi*) in the Blue Ridge region of Virginia, North Carolina, and Tennessee. Stephen G. Tilley and coauthors (*Zoological Journal of the Linnean Society* 152:115–130, 2008) clearly demonstrated the dividing line between Northern Dusky Salamanders (*Desmognathus fuscus*) and Flat-headed Salamanders (*D. planiceps*) in Virginia and North Carolina and Shawn R. Kuchta (*Herpetologica* 63:332–350, 2007) detailed the contact zone centered on the Kaweah River between California Newts (*Taricha torosa*) and Sierra Newts (*T. sierrae*) in the southern Sierra Nevada of California. Evon R. Hekkala and coauthors (*Conservation Genetics* 12:1379–1385, 2011) used DNA sequence data from museum specimens to show that frogs from near the Mogollon Rim in central Arizona previously thought to be Chiricahua Leopard Frogs (*Lithobates chiricahuensis*) appear actually to be Vegas Valley Leopard Frogs (*L. fisheri*), which had been thought to be extinct. Our maps reflect these discoveries.

Other decisions we have made about placing range boundaries between recently recognized species have not necessarily been so easy. In these cases, we have mapped the ranges according to available information and our own understanding of amphibian biogeography. These decisions are, effectively, hypotheses of the ranges of these species. Additional investigations ultimately will tell how accurate our maps may be. For example, Pacific Treefrogs were previously thought to be a single species distributed over a large part of western North America from southern British Columbia to northern Mexico. Ernesto Recuero and coauthors (*Molecular Phylogenetics and Evolution* 39:293–304, 2006) have shown, however, that there are actually three closely similar species now known as Pacific Treefrogs (*Pseudacris regilla*), Sierran Treefrogs (*P. sierrae*), and Baja California Treefrogs (*P. hypochondriaca*), all apparently with mutually exclusive ranges. They deserve separate maps, but scant information was provided concerning the precise range boundaries between these species in British Columbia and adjacent U.S. states; therefore, we have based our maps partly on our understanding of the distributions of other amphibian species in the same region.

The distribution of species among chorus frogs has also been subject to extensive investigations. James Platz (*Copeia* 1989:704–712, 1989) presented an analysis of the morphologic and mating-call differences between Boreal Chorus Frogs and Western Chorus Frogs and elevated them to full species status as *Pseudacris maculata* and *P. triseriata*, respectively. More recently, Emily Moriarty Lemmon and coauthors (*Molecular Phylogenetics and Evolution* 44:1068–1082, 2007) used mitochondrial DNA variation and phylogenetic analysis to map genetic divisions between these and six other species of closely related chorus frogs. In doing so, these authors found that the chorus frogs in southern Quebec, most of southern Ontario and northern New York, though recognized elsewhere as Western Chorus Frogs, possessed mitochondrial DNA like that of Boreal Chorus Frogs. However, here we have conservatively retained these as Western Chorus Frogs, since there are large gaps in the range between these populations and the known range limit of Boreal Chorus Frogs in northern Ontario, the James Bay region of Quebec, and the Upper Peninsula of Michigan, and because mitochondria are known to cross species boundaries between closely related species. The extent of genetic discordance between animal mitochondrial and nuclear genomes has been summarized by David P.L. Toews and Alan Brelsford (*Molecular Ecology* 21: 3907–3930, 2012). Mitochondrial–nuclear DNA discordance has been demonstrated by Brian E. Fontenot and coauthors (*Molecular Phylogenetics and Evolution* 59:66–80, 2011) among species of *Anaxyrus* toads and by John J. Wiens and coauthors (*Evolution* 60:2585–2603, 2006) among species of *Plethodon* salamanders. We expect that additional investigations will be needed before the identities of chorus frogs are fully clarified.

Two recently identified species are not included in this book as they are currently unnamed. Gregory B. Pauly (Ph.D. dissertation, University of Texas, Austin, 2008) investigated variation in Western Toads and showed extensive differences between populations (located largely in Alberta east of the Rocky Mountains) in which the males call and populations elsewhere in the range in which the males do not call. The calling populations evidently comprise a distinct species. Similarly, Catherine E. Newman and coauthors (*Molecular Phylogenetics and Evolution* 63:445–455, 2012) have found another species related to Pickerel Frogs in southern New York and northern New Jersey. The extent of the ranges of these apparently new species is not fully known.

SOURCES OF ERROR

Mapping errors are inescapable; valid records may be missed, inaccurate records may be mapped, and it is extremely difficult to correct them all. The maps we generated were based on well over a million species occurrence records, and we inspected them carefully to detect possible errors. We deleted records we thought to be questionable. This was necessarily a qualitative exercise, and the maps should not be used for determining definitive species ranges, but rather treated as general distribution guidelines in a biogeographical context. Since map units and grid cells usually included multiple and/or duplicate records, only units and cells in which all records were questionable were excluded from the final maps. U.S. maps based on county occurrences have been available online on both the USGS and AmphibiaWeb websites since 2005 and have therefore been scrutinized over the past 7 years, allowing us to eliminate numerous identified errors from the U.S. distributions as well as incorporate newly discovered range extensions and taxonomic changes. However, this book marks the first time since E.B.S (Shelley) Logier and Clive C. Toner's 1961 *Check list of the Amphibians and Reptiles of Canada and Alaska* (Royal Ontario Museum, Toronto) that all species of amphibians occurring in Canada have been mapped using dot maps.

Because mapping occurrence records is not a simple thing, even for very well-known and easily recognizable species, future corrections inevitably will be made. Data reliability is always a concern. Museum records are usually the most reliable source of accurate data, as they are frequently collected by knowledgeable professionals and are backed up by actual specimens that can be examined and verified for proper identification. But even museum records can be in error, as species may be misidentified, data may be transcribed inaccurately, or records may not be updated with the newest taxonomic information. The monumental scope of the dataset that went into making this book meant that we could not personally verify all questionable museum records or examine all specimens of interest. In some instances, museum staff examined and verified such specimens for us but, in general, we chose to be cautious and left questionable records unmapped even where this had the effect of eliminating a dot on our map. Data derived from observations, survey programs, and field notes for which there are no associated specimens almost always cannot be verified after the fact. These records were considered carefully and skeptically, especially in cases in which they appeared to indicate significant range extensions.

Even assuming that species are accurately identified, there are many sources for geo-referencing errors that may place an observation at the wrong spot on the map. Geographical coordinates may be written down incorrectly in notebooks or database records. Older records taken before global positioning satellites came into wide use usually do not have latitude and longitude or universal transverse Mercator (UTM) coordinates and instead note only distances from crossroads or particular landmarks. But roads may be rebuilt, small towns and villages may merge or disappear, sites may change, and multiple localities may sport the same name. It becomes an understandable challenge to verify and map a record for "Mud Lake," for instance, considering that there are some 248 Mud Lakes in Canada and about 860 in the U.S.

We invite corrections, and we encourage readers to contribute new records with detailed information and descriptions of the animals that are as accurate as possible. Whenever possible, voucher specimens or photographs should be deposited in public institutions, as these are the best possible evidence of the occurrences of species. Such records are valuable for the present and for the future, when amphibian distributions may be quite different from what we see today. In this way, anyone can make a valuable contribution to our better understanding of North American amphibians.

FURTHER READING

NORTH AMERICAN AMPHIBIANS

Church, D., D.M. Green, G. Hammerson, J. Mitchell, G. Parra Olea, and G. Santos Barrerra. 2008. Amphibians of the Nearctic Region. Pp. 85–92. *In* S.N. Stuart, M. Hoffmann, J.S. Chanson, N.A. Cox, R.J. Berridge, P. Ramani, and B.E. Young (Eds.), Threatened Amphibians of the World. Lynx Ediciones, Madrid.

Crother, B.I. (Ed.). 2012. Scientific and standard English names of amphibians and reptiles of North America north of Mexico. 7th edition., SSAR Herpetological Circulars 39. Society for the Study of Amphibians and Reptiles, St. Louis.

Dodd, C.K., jr. 2013. Frogs of the United States and Canada. Johns Hopkins University Press, Baltimore.

Duellman, W.E., and S.S. Sweet. 1999. Distribution patterns of amphibians in the Nearctic region of North America. Pp. 31–109. *In* W.E. Duellman (Ed.), Patterns of Distribution of Amphibians: A Global Perspective. Johns Hopkins University Press, Baltimore.

Frost, D.R., T. Grant, J. Faivovich, R. Bain, A. Haas, C.F.B. Haddad, R. de Sá, A. Channing, M. Wilkinson, S.C. Donnellan, C. Raxworthy, J.A. Campbell, B.L. Blotto, P. Moler, R.C. Drewes, R.A. Nussbaum, J.D. Lynch, D.M. Green, and W. Wheeler. 2006. The amphibian tree of life. Bulletin of the American Museum of Natural History 297:1–370

Green, D.M. (Ed.). 1997. Amphibians in Decline. Canadian Studies of a Global Problem. Society for the Study of Amphibians and Reptiles, St. Louis.

Lannoo M.J. (Ed.). 2005. Amphibian declines: The Conservation Status of United States Species. University of California Press, Berkeley.

Petranka, J.W. 1998. Salamanders of the United States and Canada. Smithsonian Institution, Washington, DC.

Stebbins, R. C., and N. W. Cohen. 1995. A Natural History of Amphibians. Princeton University Press, Princeton, NJ.

Wells, K.D. 2007. The Ecology and Behavior of Amphibians. University of Chicago Press, Chicago.

In addition to the works listed above, there are numerous provincial, state, and regional field guides, handbooks, and websites devoted to amphibians in North America. All of them contain illustrations, maps, and information about the species in the areas they cover.

For information specifically on Canadian species of amphibians, readers may consult the species status reports produced by the Committee on the Status of Endangered Wildlife in Canada

(COSEWIC), which are available from the COSEWIC Secretariat, c/o the Canadian Wildlife Service, Ottawa, or at www.cosewic.gc.ca.

NORTH AMERICAN GEOGRAPHY AND GEOLOGICAL HISTORY

Flannery, T. 2001. The Eternal Frontier: An Ecological History of North America and Its Peoples. Atlantic Monthly Press, New York.

King, P.B. 1977. The Evolution of North America. 2nd edition. Princeton University Press, Princeton, NJ.

McKnight, T.L. 2003. Regional Geography of the United States and Canada. 4th edition. Prentice Hall, Upper Saddle River, NJ.

North America. (2013). In Encyclopædia Britannica. Available from http://www.britannica.com/EBchecked/topic/418612/North-America

Sisson, V.B. 2003. The Geology of North America. *In* K. Stüwe and B. Grasemann (Eds.), Earth and Atmospheric Science, Geology (EOLSS theme 6.15. topic 2 and 7), Encyclopedia of Life Support Systems. EOLSS Publishers, Oxford, UK. Available at http://www.eolss.net/E6-15-toc.aspx.

Stearn, C.W., R.L. Carroll, and T.H. Clark. 1979. Geological Evolution of North America, 3rd edition. Wiley, New York.

NOTES

1. (page x) Some notable changes to the scientific names of species, particularly genera, compared to those used in Lannoo's 2005 book *Amphibian Declines* include reclassifying North American bufonid toads from the genus *Bufo* to the genera *Anaxyrus*, *Incillius* or *Rhinella*, moving the majority of North American ranid frogs from *Rana* to *Lithobates*, subsuming *Pternohyla* into *Smilisca*, and moving the Barking frog from the genus *Eleutherodactylus* to *Craugastor*. However, the common names for these species are largely unchanged. See the 7th edition of *Scientific and standard English names of amphibians and reptiles of North America north of Mexico*, edited by Brian I. Crother (*SSAR Herpetological Circulars* 39, 2012) for notes on current nomenclature.

2. (page 48). Tony Gamble, Peter B. Berendzen, H. Bradley Shaffer, David E. Starkey and Andrew M. Simons (*Molecular Phylogenetics and Evolution* 48: 112–125, 2008) recognized *Acris blanchardi* as a species distinct from *A. crepitans*. This account is drawn from relevant portions of the account of *A. crepitans* in *Amphibian Declines*. See also work by Kaela B. Beauclerc, Bob Johnson and Bradley N. White (*Canadian Journal of Zoology* 88:553–566, 2010).

3. (page 67) Emily Moriarty Lemmon, Alan R. Lemmon, Joseph T. Collins, Julie A. Lee-Yaw, and David C. Cannatella (*Molecular Phylogenetics and Evolution* 44: 1068–1082, 2007) confirmed that *Pseudacris feriarum* was a species separate from *P. triseriata*. This account is drawn from relevant portions of the account of the *P. triseriata* complex in *Amphibian Declines*.

4. (page 68) Emily Moriarty Lemmon, Alan R. Lemmon, Joseph T. Collins, and David C. Cannatella (*Zootaxa* 1675: 1–30, 2008) described *Pseudacris fouquettei* a new species of chorus frog distinct from *P. feriarum*, with which it had previously been considered synonymous, and provided a blob map of its distribution. The species had been treated in *Amphibian Declines* as part of the *P. triseriata* complex; this account is derived from the relevant information in that account.

5. (page 69) Ernesto Recuero, Íñigo Martínez-Solano, Gabriela Parra-Olea, and Mario García-París. (*Molecular Phylogenetics and Evolution* 39: 293–304, 2006) found evidence to divide the Pacific Treefrog, *Pseudacris regilla*, into three species (*P. regilla*, *P. hypochondriaca* and *P. sierrae*) with non-overlapping ranges. Our maps of these species are based on the map they provide, with additional interpretations (see page 317). Our accounts of *P. hypochondriaca* and *P. sierrae* are based on the relevant information contained in the account of *P. regilla* in *Amphibian Declines*.

6. (page 71) Emily Moriarty Lemmon, Alan R. Lemmon, Joseph T. Collins, Julie A. Lee-Yaw, and David C. Cannatella (*Molecular Phylogenetics and Evolution* 44: 1068–1082, 2007) distinguished that *Pseudacris kalmi* from *P. triseriata* as a separate species. This account is based on the relevant content in the account of the *P. triseriata* complex in *Amphibian Declines*.

7. (page 72) *Pseudacris maculata* was distinguished as a separate species from *P. triseriata* by Emily Moriarty Lemmon, Alan R. Lemmon, Joseph T. Collins, Julie A. Lee-Yaw, and David C. Cannatella (*Molecular Phylogenetics and Evolution* 44: 1068–1082, 2007). This account is based in part on relevant portions of the account of the *P. triseriata* complex in *Amphibian Declines*. Our range map of the species, however, differs from the map based on mitochondrial DNA variation presented by Moriarty Lemmon et al. See page 317 for explanation.

8. (page 79) See page 317 for explanation on mapping *Pseudacris triseriata*.

9. (pages 100 and 102) Our maps of *Lithobates chiricahuensis* and *L. fisheri* reflect findings by Evon R. Hekkala, Raymond A. Saumure, Jef R. Jaeger, Hans-Werner Herrmann, Michael J. Sredl, David F. Bradford, Danielle Drabeck, and Michael J. Blum (*Conservation Genetics* 12:1379–1385, 2011) indicating that some leopard frogs in central Arizona are likely Vegas Valley Leopard Frogs rather than Chiricahua Leopard Frogs. See pages 316–317.

10. (page 111). Catherine E. Newman, Jeremy A. Feinberg, Leslie J. Rissler, Joanna Burger, and H. Bradley Shaffer. (*Molecular Phylogenetics and Evolution* 63: 445–455, 2012) show that leopard frogs in the extreme northeast of the range of *Lithobates sphenocephalus* are actually another species, as yet un-named. See page 318.

11. (page 135). Julie A. Lee-Yaw and Darren E. Irwin (*Journal of Evolutionary Biology* 25: 2276–2287, 2012) have identified multiple lineages of Long-toed Salamanders distinct enough that they might be considered separate species. However, these have not as yet been formally recognized and their geographic boundaries remain largely undefined.

12. (page 142) See papers by James P. Bogart, Ke Bi, Jinzong Fu, Daniel W.A. Noble, and John Niedzwiecki (*Genome* 50: 119–136, 2007) and Ke Bi and James P. Bogart (*BMC Evolutionary Biology* 10:238) for further information.

13. (pages 157 and 159) Elizabeth L. Jockusch, Íñigo Martinez-Solano, Robert W. Hansen, and David B. Wake (*Zootaxa* 3190: 1–30, 2012) described the new species, *Batrachoseps altasierrae* and *B. bramei*, for populations of slender salamanders from the southern Sierra Nevada that had previously been thought to be the Relictual Slender Salamanders, *B. relictus* (page 173). Our accounts of these two new species are based on applicable information in the account of *B. relictus* in *Amphibian Declines*.

14. (Page 193) Erica J. Crespi, Robert A. Browne, and Leslie J. Rissler (*Herpetologica* 66:283–295, 2010) described *Desmognathus organi* and delineated the boundary of its range relative to its sister species, *D. wrighti* (page 198). Our account of this species account is based on this work and relevant details in the account of *D. wrighti* in *Amphibian Declines*.

15. (page 240) Louise S. Mead, David R. Clayton, Richard S. Nauman, Deanna H. Olson and Michael E. Pfrender (*Herpetologica* 61:158–177, 2005) described *Plethodon asupak* and demonstrated its distinction from *P. stormi*, with which it has previously been confused. The ecology and conservation status of both species were reviewed by Douglas J. DeGross and R. Bruce Bury (*U.S. Geological Survey, Reston, Virginia*, Open-File Report 2007–1352, 2007)

16. (page 294) *Urspelerpes brucei* was described by Carlos D. Camp, William E. Peterman, Joseph R. Milanovich, Trip Lamb, John C. Maerz, and David B. Wake (*Journal of Zoology* 279:86–94, 2009), who recount its discovery and provide notes on its ecology and life history.

17. (page 309) Shawn R. Kuchta (*Herpetologica* 63:332–350, 2007) concluded that newts in the northern Sierra Nevada were a species distinct from the California Newts, *Taricha torosa*, in the southern Sierra Nevada. He therefore recognized them as *T. sierrae* and detailed the contact zone between the two species. Our account of *T. sierrae* is based on the relevant information from the account of *T. torosa* in *Amphibian Declines*.

18. (page 313) We follow the 7th edition of *Scientific and standard English names of amphibians and reptiles of North America north of Mexico*, edited by Brian I. Crother (*SSAR Herpetological*

Circulars 39, 2012), and Alain Dubois and Jean Raffaëlli (*Alytes* 28: 77–161, 2012) in treating the sirens in the Rio Grande region of southern Texas as *Siren intermedia*, although they were accorded a separate account in *Amphibian Declines*. As explained by Darrel R. Frost and Michael J. Lannoo (p. 914 in *Amphibian Declines*), whether these salamanders are *S. intermedia*, *S. lacertina*, or something else completely ("*S. texana*") is unresolved.

19. (page 315) The use of product names does not imply endorsement by the U.S. Federal government.

ACKNOWLEDGEMENTS

An undertaking of this scope requires the assistance and cooperation of a great many people. We are hugely indebted to the many persons and institutions that provided data for informing the maps. We particularly thank Sara Lourie, Jay Ploss, Mike Jones, and Nathan Engbrecht for their assistance in finding and georeferencing Canadian locality data, Stefane M. Nadeau for assistance with data acquisition and management, Kinard Boone, Lynda Garrett, Kimberly Gazenski, Baboyma Kagniniwa, Allison Sussman, and Mark Wimer for technical assistance, and Evan Grant and Cynthia Paszkowski for comments on the text. We acknowledge the following people for their assistance in maintaining the U.S. data as part of the U.S. Geological Survey National Amphibian Atlas: Ovais Aslam, Jang Byun, Kevin Laurent, Melissa Miller, Crystalina McGrail, Bruce Peterjohn, Jessica Sushinsky, and Mark Wimer. We thank the following providers of data: Academy of Natural Sciences, Philadelphia (Ned S. Gilmore); Alberta Natural Heritage Information Centre, Edmonton, AB (Wayne Nordstrom, Drajs Vujnovic, Lorna Allen); American Museum of Natural History, New York; Amphibian and Reptile Collection, University of Arizona, Tucson; Amphibian and Reptile Diversity Research Center, University of Texas at Arlington; Arkansas State University, State University (Stan Trauth); Atlantic Canada Conservation Data Centre, Sackville, NB (R.A. Lautenschlager); Auburn University Natural History Museum and Learning Center, Auburn, AL; Austin Peay State University, Clarksville, TN (A. Floyd Scott); Bernice P. Bishop Museum, Honolulu; Borror Laboratory of Bioacoustics, Ohio State University, Columbus; British Columbia Conservation Data Centre, Ministry of Environment, Victoria (Erin Prescott, Gail Harcombe, Marta Donovan); British Columbia Ministry of Water, Land, and Air Protection, Victoria (Laura Friis); California Academy of Sciences, San Francisco (Jens Vindum); Canadian Museum of Nature, Ottawa, ON (Michele Steigerwald, Peter Frank, Francis R. Cook, Frederick W. Scheuler); Carnegie Museum of Natural History, Pittsburgh (Charles McCoy); Carolina Herp Atlas, Davidson College, Davidson, NC (Steven J. Price); Committee on the Status of Endangered Wildlife in Canada, Ottawa, ON (Shirley Hamelin, Alain Filion); Cornell University Museum of Vertebrates, Ithaca, NY; Cowan Vertebrate Museum, University of British Columbia, Vancouver (Nasar Din, Rex D. Kenner); David Goodward; Environment and Natural Resources, Government of the Northwest Territories (Suzanne Carrière); Field Museum of Natural History, Chicago (Alan Resetar, Harold K. Voris); Fish and Wildlife Management Division, Alberta Sustainable Resource Development, Edmonton (Lonnie Bilyk); Florida Museum of Natural History, Gainesville (Max Nickerson, Kenneth L. Krysko); FrogWatch, Nature Canada, Ottawa, ON; Georgia Department of Natural Resources (John Jensen); Grand Canyon National Park, Grand Canyon, AZ (Colleen L. Hyde); Illinois Natural History Survey, University of Illinois, Champaign (Chris Phillips, Chris

Mayer); Indiana Department of Natural Resources, Indianapolis (Lee Casebere); Iowa Department of Natural Resources, Boone (Stephanie Shepherd); James Ford Bell Museum of Natural History, Minneapolis (Kenneth Kozak, Christopher E. Smith); Kentucky Department of Fish and Wildlife Resources, Frankfort (John MacGregor); Los Angeles County Museum of Natural History, Los Angeles (J. Wright); Louisiana Department of Wildlife and Fisheries, Baton Rouge (Jeff Boundy); Louisiana Museum of Natural History, Louisiana State University, Baton Rouge; Manitoba Herps Atlas, Winnipeg (Doug Collicutt); Manitoba Museum, Winnipeg (Randall Mooi); Manitoba Wildlife and Ecosystem Protection Branch, Manitoba Conservation, Winnipeg (Nicole Firlotte, Bill Watkins); Marshall University (Tom Pauley); Michigan State University Museum, East Lansing (Jim Harding); Milwaukee Public Museum, Milwaukee, WI; Ministère des Ressources Naturelles et de la Faune, Québec (Annie Paquet); Minnesota Department of Natural Resources, St. Paul (Jeffrey LeClere); Mississippi Museum of Natural Science, Jackson (Tom Mann); Montana Natural Heritage Program, Helena (Bryce Maxell); Monte L. Bean Museum, Brigham Young University, Provo, UT; Museum of Comparative Zoology, Harvard University, Cambridge, MA; Museum of Natural History, Santa Barbara, CA (Krista A. Fahy); Museum of Natural History, University of Illinois, Urbana (Donald I. Hoffmeister); Museum of Northern Arizona, Flagstaff (Janet Whitmore Gillette); Museum of Southwestern Biology, University of New Mexico, Albuquerque (J. Tom Giermakowski); Museum of Vertebrate Zoology, University of California, Berkeley (David Wake, Stephen D. Busack); Museum of Zoology, University of Calgary, Calgary, AB (Barry Curtis); Museum of Zoology, University of Michigan, Ann Arbor (Greg Schneider); National Museum of Natural History, Smithsonian Institution, Washington, DC (Kevin de Queiroz); Natural History Museum of Los Angeles County, Los Angeles; Natural History Museum, San Diego (Mark Dodero); Natural Resources Research Institute, University of Minnesota Duluth (Lucinda B. Johnson); NatureServe, Arlington, VA (Lynn Kutner, Leslie Honey); New Brunswick Museum, Frederickton (Don McAlpine); North Carolina State Museum of Natural Sciences, Raleigh (Bryan Stuart); North Carolina Wildlife Resources Commission, Raleigh (Jeff Humphries); Ontario Herpetofaunal Summary, Ministry of Natural Resources, Peterborough (Michael Oldham); Partners in Amphibian and Reptile Conservation, Washington, DC (Priya Nanjappa); Peabody Museum, Yale University, New Haven, CT; Provincial Museum of Alberta, Edmonton (H.C. Smith); Redpath Museum, McGill University, Montreal, QC; Redwood Sciences Lab, Arcata, CA (Hartwell H. Welsh, Jr.); Royal British Columbia Museum, Victoria (James A. Cosgrove); Royal Ontario Museum, Toronto (James Lovisek, Ross MacCulloch); Sam Noble Oklahoma Museum, University of Oklahoma, Norman (Janalee Caldwell, Jessica Watters); San Diego Natural History Museum, San Diego; Santa Barbara Natural History Museum, Santa Barbara, CA; Saskatchewan Conservation Data Centre, Saskatchewan Environment, Regina (Ben Sawa); Saskatchewan Herpetology Atlas Project, Saskatoon (Andrew Didiuk); Savannah River Ecology Lab, Aiken, SC (Thomas Luhring); Société d'Histoire Naturelle de la Vallée du Saint-Laurent, Ste-Anne-de-Bellevue, QC (David Rodrigue, Sébastien Rouleau); South Dakota Department of Game Fish and Parks, Pierre (Doug Backlund); Staatliches Museum fur Naturkunde Stuttgart, Germany; Sternberg Museum of Natural History, Hays, KS (Travis W. Taggart); Texas Cooperative Wildlife Collection, Texas A & M University, College Station (James R. Dixon); Texas Natural History Center, University of Texas, Austin; Texas State University, San Marcos (Michael Forstner); The Nature Conservancy, Boulder, CO (Chris Pague); U.S. Fish and Wildlife Service, Tucson, AZ (Jim Rorabaugh); U.S. Geological Survey National Wetlands Research Center, Lafayette, LA (Hardin Waddle); U.S. Geological Survey Patuxent Wildlife Research Center, Laurel, MD (Eric Dallalio); U.S. Geological Survey Southeast Ecological Science Center, Gainesville, FL (William J. Barichivich); U.S. Geological Survey, Sioux Falls, SD (Alisa Gallant); University of Alabama Museum of Natural History, Tuscaloosa; University of Alaska Museum of the North, Fairbanks; University of Alberta, Edmonton (Cynthia Paszkowski, Jim Whittome, Arthur Whiting); University of Colorado Museum of Natural History, Boulder

(Mariko Kageyama); University of Illinois Museum of Natural History, Champaign (Chris Phillips, Chris Mayer); University of Kansas Natural History Museum and Biodiversity Research Center, Lawrence (Andrew Campbell, Rafe M. Brown); University of Louisiana, Monroe (John L. Carr); University of Nebraska State Museum, Lincoln (Thomas E. Labedz); University of Wisconsin, Madison (Laura Halverson); University of Wisconsin, Stevens Point (Erik Wild); Utah Museum of Natural History, Salt Lake City; Virginia Herpetological Society, Newport News (Jason Gibson); Wisconsin Herp Atlas, University of Wisconsin, Madison Field Station, Saukville; Yukon Department of Environment, Whitehorse (Brian G. Slough, Syd Cannings, Tom Jung); Zoological Institute, Russian Academy of Sciences, St. Petersburg, and all the many, many people who have provided records of amphibians in North America to these agencies, programs, and institutions.

We thank the following authors of the original species accounts: Michael J. Adams, Carl D. Anthony, Mark A. Bailey, Christopher K. Beachy, David A. Beamer, Laura Blackburn, Sean M. Blomquist, Ronald M. Bonett, Jeff Boundy, David F. Bradford, Alvin L. Braswell, Robert Brodman, Ronald A. Brandon, Lauren E. Brown, R. Bruce Bury, Brian P. Butterfield, Carlos D. Camp, John Cavagnaro, Paul T. Chippindale, George R. Cline, Christopher Conner, Paul Stephen Corn, John A. Crawford, John J. Crayon, Carlos Davidson, Gage H. Dayton, C. Kenneth Dodd Jr., Brooke A. Douthitt, Darrin Doyle, Harold A. Dundee, Edward L. Ervin, Michael A. Ewert, Eugenia Farrar, Zachary I. Felix, Gary M. Fellers, Dante B. Fenolio, Kimberleigh J. Field, M. J. Fouquette Jr., Joe N. Fries, Julie A. Fronzuto, Erik W.A. Gergus, Anna Goebel, Caren S. Goldberg, Robert H. Goodman Jr., Brent M. Graves, Robert H. Gray, Craig Guyer, Stephen F. Hale, Robert Hansen, Robert W. Hansen, Reid N. Harris, Julian R. Harrison, Marc P. Hayes, Russ Hendricks, Jean-Marc Hero, Robert E. Herrington, Jane Hey, W. Ronald Heyer, W. Jeffrey Humphries, Todd W. Hunsinger, Kelly J. Irwin, Jef R. Jaeger, Randy D. Jennings, John B. Jensen, Steve A. Johnson, Lawrence L.C. Jones, Frank W. Judd, J. Eric Juterbock, Nancy E. Karraker, Michelle S. Koo, Kenneth H. Kozak, Fred Kraus, James J. Krupa, Shawn R. Kuchta, James Lazell, William T. Leja, Kirk Lohman, Sharyn B. Marks, Timothy O. Matson, D. Bruce Means, Joseph R. Mendelson III, Walter E. Meshaka Jr., Joseph C. Mitchell, John H. Malone, Paul E. Moler, David J. Morafka, Steven R. Morey, Emily Moriarty, Jennifer Mui, Robert W. Murphy, Erin Muths, Priya Nanjappa, R. Andrew Odum, Deanna H. Olson, Richard B. Owen, Charles W. Painter, John G. Palis, Theodore J. Papenfuss, Duncan Parks, Matthew J. Parris, Beth Anne Pauley, Thomas K. Pauley, Christopher A. Pearl, William Peterman, Christopher A. Phillips, David S. Pilliod, Fred Punzo, Timothy Quinn, Cindy Ramotnik, Jamie K. Reaser, Michael Redmer, Stephen C. Richter, James C. Rorabaugh, Travis J. Ryan, Wesley K. Savage, Cecil R. Schwalbe, Terry D. Schwaner, David E. Scott, David M. Sever, H. Bradley Shaffer, Donald B. Shepard, Hobart M. Smith, Michael J. Sredl, Nancy L. Staub, Margaret M. Stewart, Melody Stoneham, Brian K. Sullivan, Samuel S. Sweet, Stephen G. Tilley, Stanley E. Trauth, Peter C. Trenham, Vance Vredenburg, David B. Wake, J. Eric Wallace, Mark B. Watson, Hartwell H. Welsh Jr., Jill A. Wicknick, and Kelly R. Zamudio.

We thank the following people for providing photographs used in this volume: Esteban Alzate, Sarah Armstrong, Richard D. Bartlett, Steve Bennett, Constance Brown, Travis Brown, Tim Burkhardt, Bruce Christman, John Clare, Erica Crespi, Paul Crump, Ken Dodd, Dante Fenolio, Dino Ferri, Pierre Fidenci, Tony Gamble, Patrick Gault, Scott Gillingwater, Brad Glorioso, Caren Goldberg, Michael Graziano, David M. Green, Joyce Gross, Nathan Haislip, Jim Harding, Darlene Hecnar, Tim Herman, Pierson Hill, Randy Jennings, Joyce Marie Klaus, Aubrey Huepel, William Leonard, Patrick Moldowan, Kevin Messenger, Brad Moon, Gary Nafis, Kristiina Ovaska, Bill Peterman, Todd Pierson, Alberto Puente-Rolon, Mike Redmer, Bruce Taubert, Steve Tilley, Jonathan Woodward, and Wolfgang Wuster. We thank Ron Blakey (Colorado Plateau Geosystems, Flagstaff, AZ) for permission to reprint the map of Pleistocene North America.

Finally, we sincerely thank Chuck Crumly, who not only initiated this project but also saw it through much of its development, and Lynn Meinhardt, who guided its assembly.

INDEX